GYOTAKU PRINTS OF FISH AND CRUSTACEANS IN SOUTHEAST ALASKA

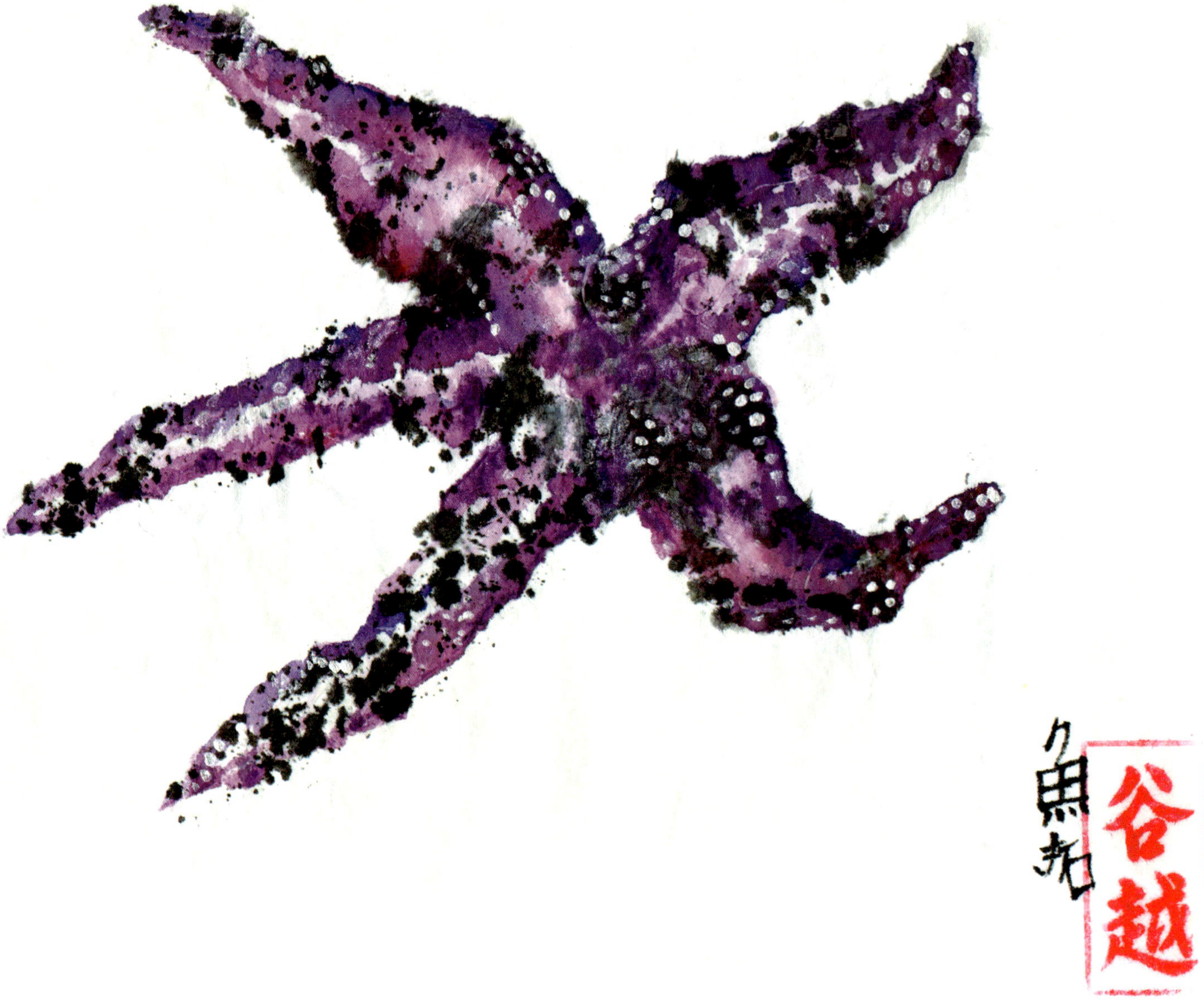
魚拓
谷越

GYOTAKU PRINTS OF FISH AND CRUSTACEANS IN SOUTHEAST ALASKA

BY JULIA TINKER

UNIVERSITY OF ALASKA PRESS FAIRBANKS

Published by University of Alaska Press
P.O. Box 756240
Fairbanks, AK 99775-6240

Cover and interior design by Kristina Kachele Design, llc

Cover image: Shiner perch, gyotaku print by Julia Tinker.

Interior photos and images by Julia Tinker.

Library of Congress Cataloging in Publication Data
Names: Tinker, Julia, author.
Title: Gyotaku prints of fish and crustaceans in Southeast Alaska / by Julia Tinker.
Description: Fairbanks, Alaska: University of Alaska Press, 2019.

Identifiers: LCCN 2018030505 | ISBN 9781602233782 (pbk.: alk. paper)

Subjects: LCSH: Tinker, Julia | Fish prints | Marine animals in art |
Fishes Alaska, Southeast | Crustacea Alaska, Southeast.

Classification: LCC NE1341.U64 T562 2019 | DDC 769.4/37dc23
LC record available at https://lccn.loc.gov/2018030505

Printed in Canada

To my grandma, Misao Tanigoshi (みさお　谷越) for patiently teaching me to love the ocean.

CONTENTS

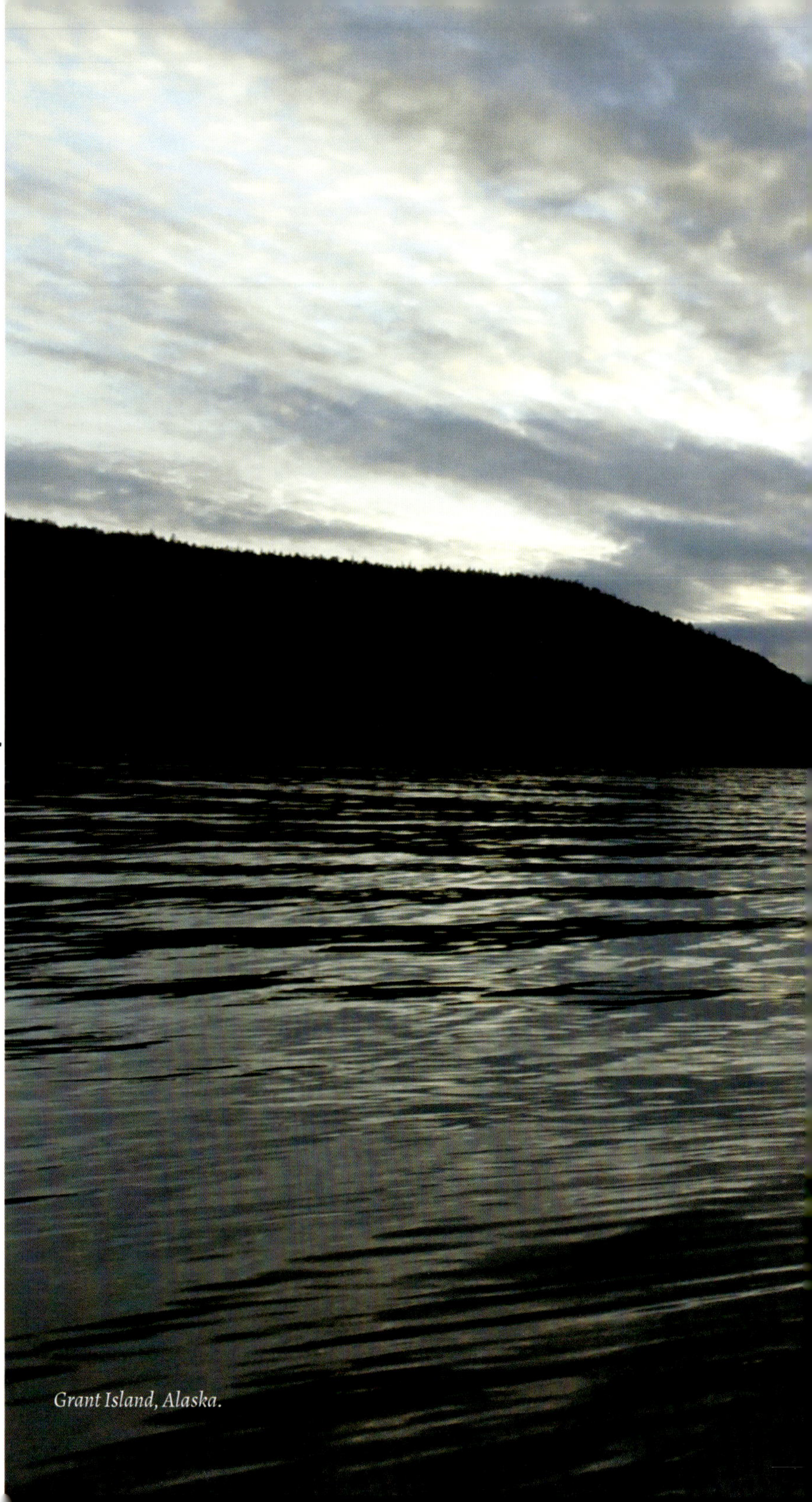

Grant Island, Alaska.

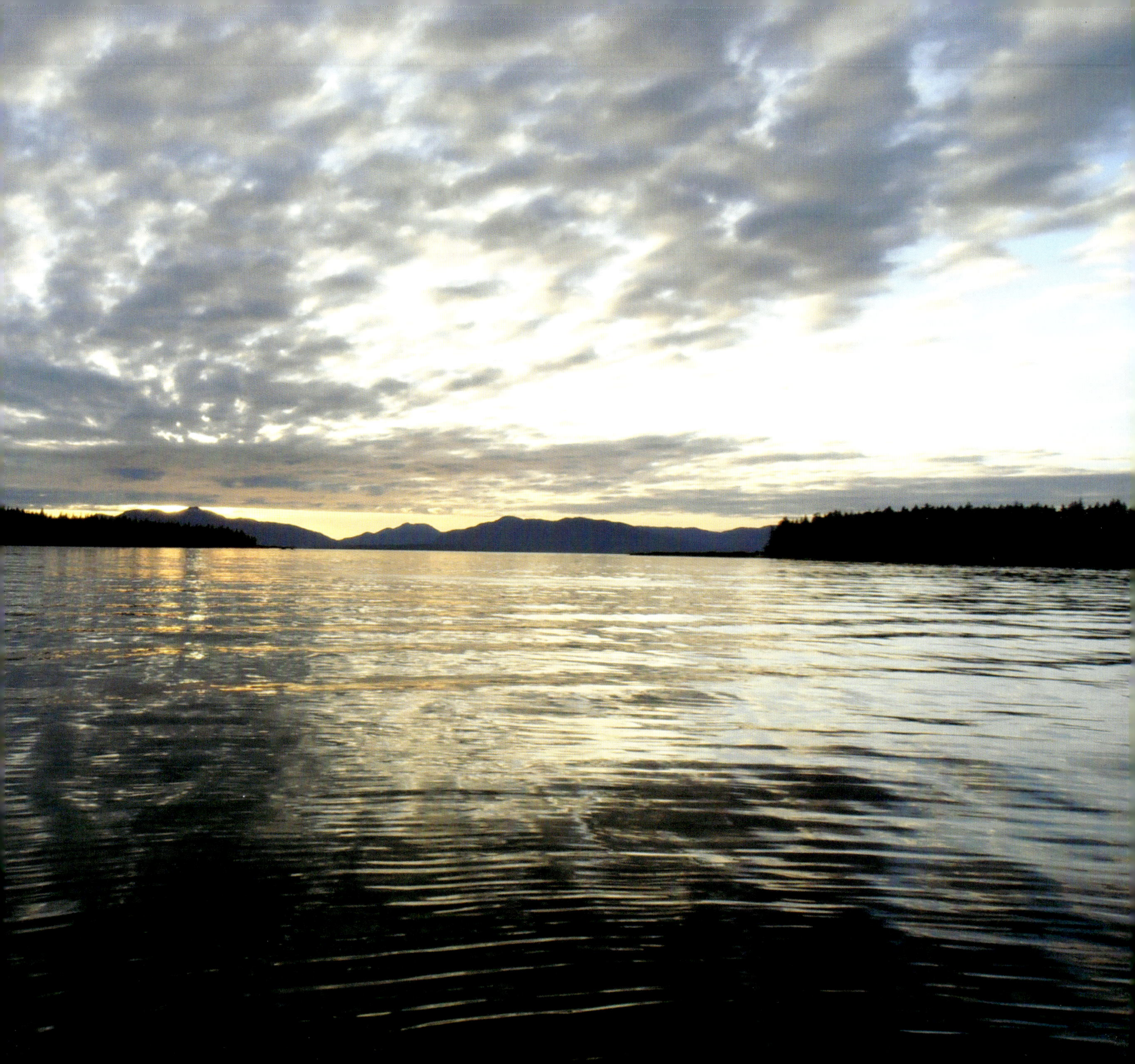

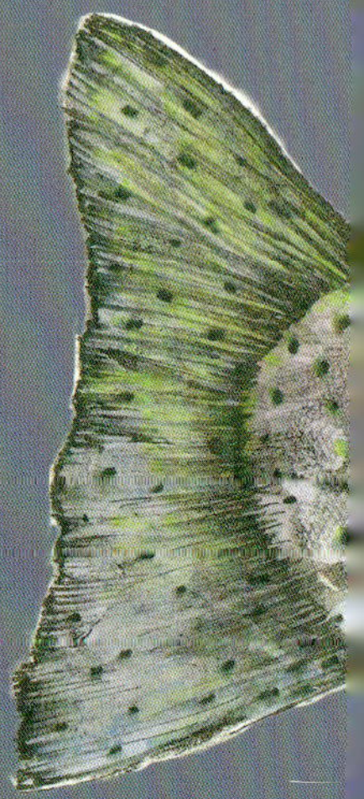

King salmon.

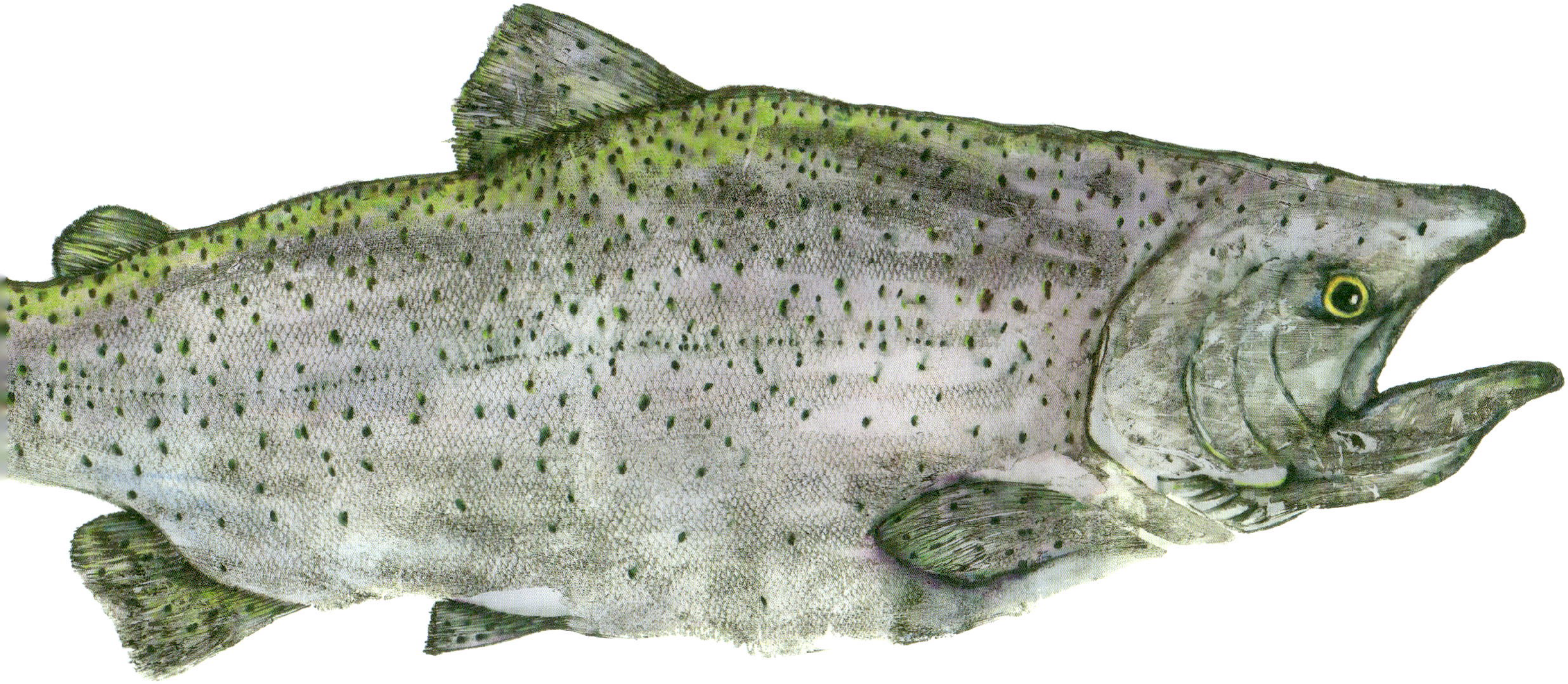

GYOTAKU PRINTS OF FISH AND CRUSTACEANS IN SOUTHEAST ALASKA

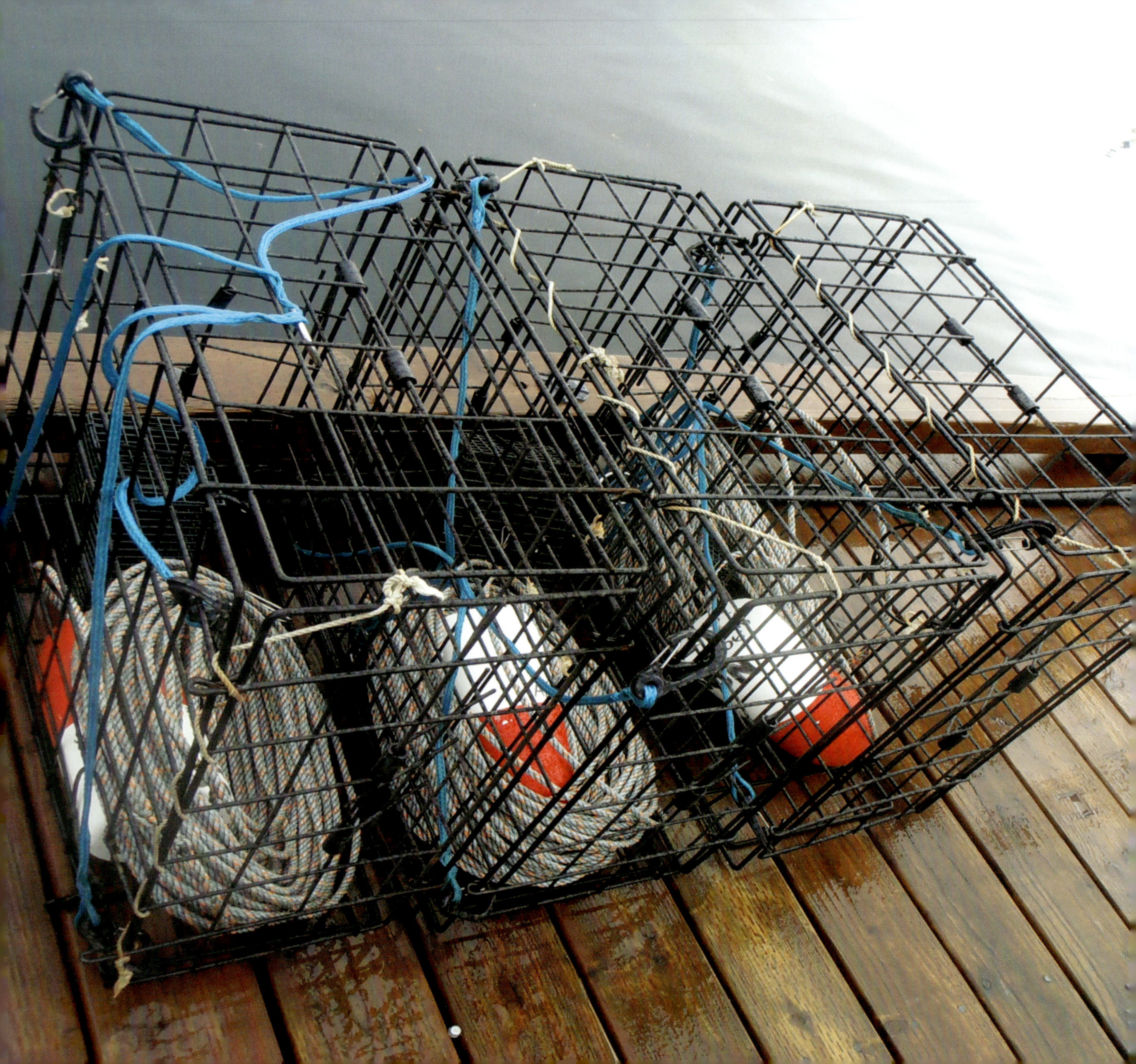

INTRODUCTION

Do you *gyotaku*? Do you look at fish prints and ask, "How did they do that?" If so, I share your excitement and curiosity. The art of gyotaku has fascinated me for as long as I can remember. All of the prints and photographs in this book represent my love for Alaska and my adoration of ancient Japanese gyotaku art. I hope you will leaf through these pages and travel with me through Southeast Alaska.

Many years ago, I happened upon a gyotaku print of a king salmon in a bed-and-breakfast in Prince of Wales. My mouth dropped because I was impressed by the simplicity of the print and moved by its beauty. I wanted to create my own gyotaku of Southeast Alaska's marine life. I set out on a gyotaku adventure, chronicling rockfish, crustaceans, and salmon that have been staples of the Native Alaskans' diet for generations. While I have made gyotaku prints in the more monochromatic style, I chose to do the paintings in this book by adding watercolor to each of the black sumi ink pressings. In this way, I was able to achieve a colorful representation of the beauty of all of the fish and crustaceans of Southeast Alaska.

Typically, one thinks of salmon and halibut when considering the fish caught and eaten in Southeast Alaska. However, Native Alaskans fished for rockfish in winter and early spring because salmon and halibut were scarce in those months.

Crab traps on dock at Clover Pass, Ketchikan, Alaska.

They fished in safer, more sheltered waters in the cold winter months. In the spring, they were able to fish in open water and caught rockfish, halibut, and salmon. Even now, the Alaska Board of Fisheries recognizes rockfish for subsistence use in Southeast Alaska.

As an artist and avid angler, my hope in illustrating and writing this book on the fish and crustaceans of Southeast Alaska is to celebrate the diversity of marine life in this region. It is also imperative to highlight the importance of the diverse marine life to Native Alaskans.

In order to find all the species included in this book, I spent many summers fishing and exploring the waters near Ketchikan and Prince of Wales in Southeast Alaska. My mornings began as the sun was coming up. The purr of the outboard engine behind me and the smell of the ocean air enticed me to search for new fishing spots. The weather was typically in the fifties and sixties and frequently there was a light drizzle as I embarked on my days. The lush, green islands in the near distance provided a backdrop to the dark blue-green water. Usually, the surface of the water was flat that early in the morning and I had a smooth departure out to the fishing grounds about ten miles away. Every now and again, though, a humpback whale would come up for air nearby and remind me of how small I really was in my fishing boat.

Channel 16 on my radio spoke to me in its scratchy voice. What would the weather be like throughout the day? Were there any weather hazards I should be aware of, like gusty winds, high seas, or impending storms out of the southwest? My 20-foot boat was equipped with life jackets, first-aid kits, flotation cushions, a satellite GPS radio with emergency backup button, and flares. The ocean is always in charge. She determines our destiny . . . not the other way around.

If you dream about being on the water, then this is your dream. Huge fishing vessels, tenders, seiners, and long liners were my company on the water. The distant bellow of the horns of encroaching vessels coming through Clarence Straits reminded me of their presence. The tides turned twice a day, and the ocean swelled and subsided in cycles. The high-pitched screech of bald eagles high up in the trees was music to my ears. I went out on my boat for at least 20 miles every day to fish for hours, dropping my line in every depth of water trying to find the quillback, the dusky, the sculpin, and the flounder. I went farther out looking for chinook, coho, and humpies. My adventure was a quest and the Southeast Alaska ocean controlled its outcome.

Each night, I made a gyotaku print of what I caught. After I finished each print, I signed it with our family's *hanko*, or stamp, of our name "Tanigoshi" in kanji. In Japan, everyone has a personal stamp for signing important documents such as bank transactions, loan papers, and even wills. In fact, after the stamp is recognized by the Japanese government, this becomes the official family signature.

I feel fortunate to have been able to illustrate and write this book. It is my hope that these gyotaku paintings and photographs of Southeast Alaska will encourage people to preserve its marine life and the area's rich cultural heritage. After all, it is truly our Last Frontier.

THE HISTORY OF GYOTAKU

When people come back from fishing, having spent hours and even days on the water, the question inevitably arises: "How big was it?" The angler stretches out his arms as wide as he can and says, "It was this big." But who really knows if this is a fish tale?

Exaggeration is at the heart of fishing. It is the fish tale that grows over time, making the actual size of the fish immeasurable. In the early 1800s in Japan, however, anglers devised a method to document their prized catches and put to rest any doubt about the size of their catch. Because they did not have cameras to photograph their fish, Japanese anglers would take rice paper and nontoxic sumi ink on their fishing expeditions and make a print of the fish they caught while on the boat. In Japanese, *gyo* means "fish" and *taku* means "pressing." The earliest gyotaku was a black-and-white print of the fish made from nontoxic sumi ink on handmade washi paper. Traditionally, washi paper is made of mulberry tree fibers and the resulting fibrous paper is strong enough to withstand the application of water and not rip. It was not until the mid-1800s that anglers began to paint in details of the fish's eyes and add vibrancy to the black-and-white print.

By using nontoxic sumi ink, the angler could make a gyotaku impression of his prized catch while it was still alive, subsequently releasing it back into the ocean. Or, he could wash the ink off of the fish and eat the fish later. The gyotaku was an indispensable (and indisputable) method of recording his catch. This print, or gyotaku, became a permanent record of the fish. Its measurements were tangible. No longer could the angler's tale make a minnow into a whale.

The popularity of gyotaku grew during the Tokugawa or Edo period (1603–1867) in Japan under the patronage of Lord Sakai of the Tohoku region. Lord Sakai wanted to have his prized catch of a red sea bream memorialized in a gyotaku. So, he commissioned a local artist to produce the fish pressing. The identity of this artist is unknown; but what is known is that Lord Sakai was so enamored with gyotaku that he continued to fill his palace with more and more of these fish pressings.

However, the emphasis that the Tokugawa Shogunate feudal government placed on traditional Japanese arts was about to change. In 1868, the Tokugawa Shogunate (also known as the Edo Shogunate) fell to the Meiji Restoration. The Meiji Restoration brought modernization to Japan and moved Japanese society further away from its emphasis on traditional Japanese art such as gyotaku.

Recently, there has been a resurgence of gyotaku as an art form in Japan and in the United States. The last few decades have seen gyotaku reemerge as both the black-and-white documentation of the type and size of the fish caught, as well as the color embellishment of the gyotaku print by the artist. The resurgence of gyotaku as an art form also lends itself to the artist making multiple prints of the same fish in one "painting," in addition to adding prints of seaweed and algae to provide a natural setting for the fish, crustaceans, squid, and octopus.

GYOTAKU TECHNIQUES

Since the late 1800s, gyotaku has continued to regain strength. What was a stark black-and-white sumi ink pressing on washi paper has evolved into two different methods—direct (*chokusetsu-ho*) and indirect (*kansetsu-ho*).

In the indirect method of gyotaku, the artist places the fish on a stable surface and applies a very thin, wetted sheet of washi paper to the naked fish. There is no sumi ink on the fish. The artist applies the color of the fish through the washi paper by pressing color-infused silk balls called *tampo* to each part of the fish. The resulting gyotaku is ethereal and beautiful.

I use the direct method of gyotaku printing, which I have detailed below. For a fun and easy project, follow the instructions and make your own gyotaku on your next trip!

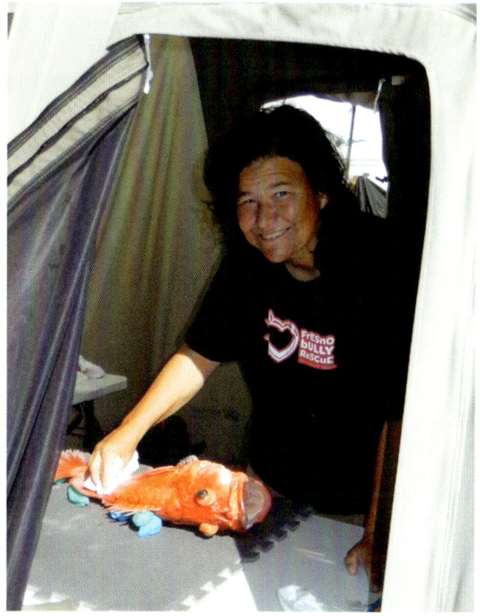

(above) Preparing the fish for printing.
(below) Applying sumi ink.

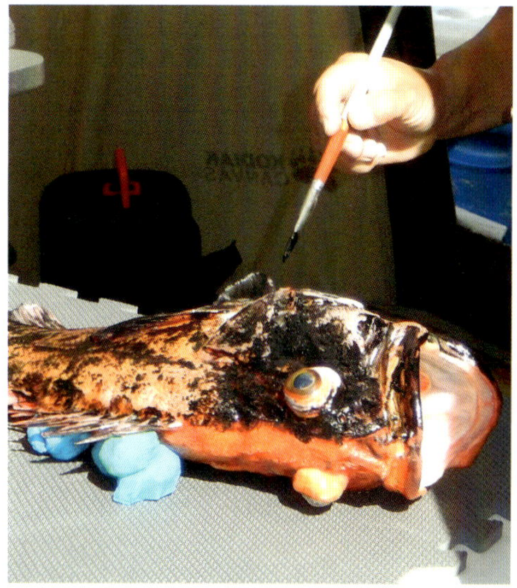

HOW TO

Step 1: Preparing the Fish for Gyotaku

First, I closely examine each side of the fish to decide which side is going to provide the cleanest and most precise mirror-image gyotaku print. When you are doing your own gyotaku, pick the side of the fish that has the most complete scales. The gyotaku copies the scales exactly as they are, so be confident that the side you pick is the side you want to print.

I keep the fish whole, but wash the outside of the fish thoroughly. To do this, I use paper towels moistened with lemon juice and salt because it cuts through the slimy film that covers the fish's scales. Carefully clean each side of your fish with the moist paper towels. I also fill the orifice under the gill plate and the anal cavity so the moisture will not bleed out onto the final print.

Step 2: Applying Sumi Ink to Fish

I use the traditional nontoxic sumi ink and a half-inch synthetic bristle paintbrush to coat the one side of the fish before applying the washi paper. Avoid painting the eye as you will come back later and complete it on your gyotaku. Nontoxic sumi ink is made from pine soot, animal fat, and water. The animal fat, or *nikawa*, has an unpleasant smell. Often incense is added to the mixture to temper the odor. You can purchase sumi ink and the washi paper at most art supply stores.

For the washi paper, it is best to get somewhere between 30 to 40 gsm, or grams per square meter. This texture will hold the ink well and will still maintain its integrity and strength when making the gyotaku.

It is important to only apply enough sumi ink so as to make a detailed mirror image of the fish, but not so much ink as to saturate the paper and create a messy blob of black paint. This takes practice, but this is the fun part!

The next step of the process must be done in one fluid step. Hold the washi paper over the inked fish and imagine where you want the print to be on the paper itself. Do you want it in the center, facing up, facing down, or to the corners and sides? When you have decided, carefully lay a piece of washi paper over the inked fish. The washi paper is strong enough so that the wetness of the ink will not cause the paper to tear as it is pressed over the jagged spines and jawline of the fish.

Removing washi paper.

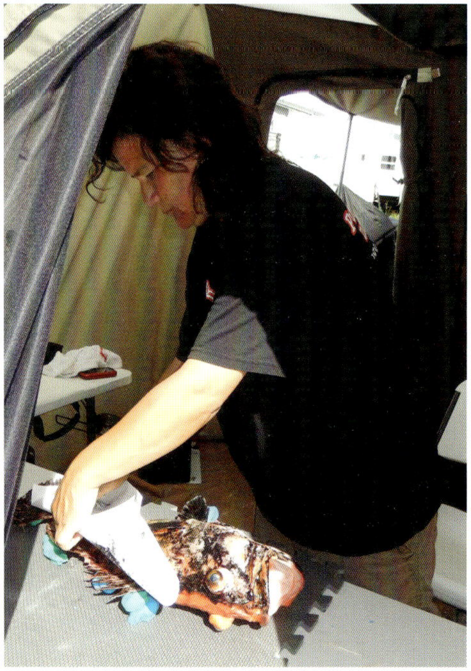

After I set the fibrous paper down, I press each part of the fish's body with my fingers. Press each part of the fish only one time in a downward, but not side-to-side motion. This step takes practice because if you accidentally press a part of the fish twice, the gyotaku will be a double print. If that happens, the mirror image of the fish will appear as two fish, or a blurred image of one fish. The gyotaku would be unintelligible. If you do this, just throw that print away, apply some more sumi ink to the fish, and start again.

It took me many years of practice to know how much ink to put on the fish and how much pressure to apply to the paper when making the gyotaku print. If you get too frustrated, put the fish back in the refrigerator for a bit and come back to it in an hour or so. You can get two to three good prints off each fish.

Step 3: Removing Washi Paper from Inked Fish

After I am confident that the fish press is complete, I gently peel off the paper in one fluid movement. The gyotaku is an exact mirror image of the fish. This is especially evident in the print of the starry flounder because its eyes look to be on its left side when, in reality, they are on its right side. Traditionally, direct method gyotaku is black and white.

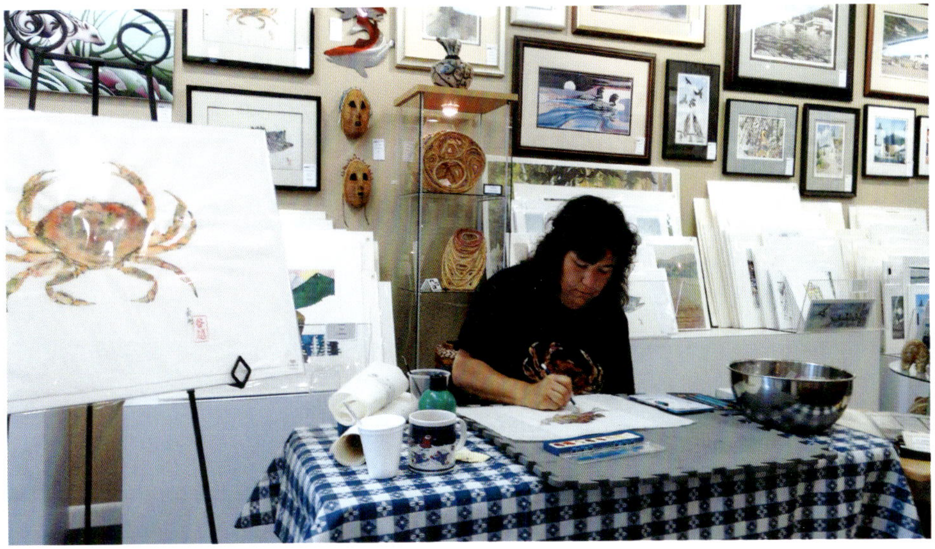

Adding watercolor to print.

Step 4: Adding to the Print

I take the black-and-white gyotaku print one step further by painting each gyotaku with watercolors and some acrylic silver paint (on the salmon scales). This is what sets my gyotaku apart from the traditional monochromatic gyotaku prints that originated in Japan in the early 1800s.

The resulting paintings are unique and represent the fish's appearance in a manner that is faithful to the ancient Japanese tradition of gyotaku.

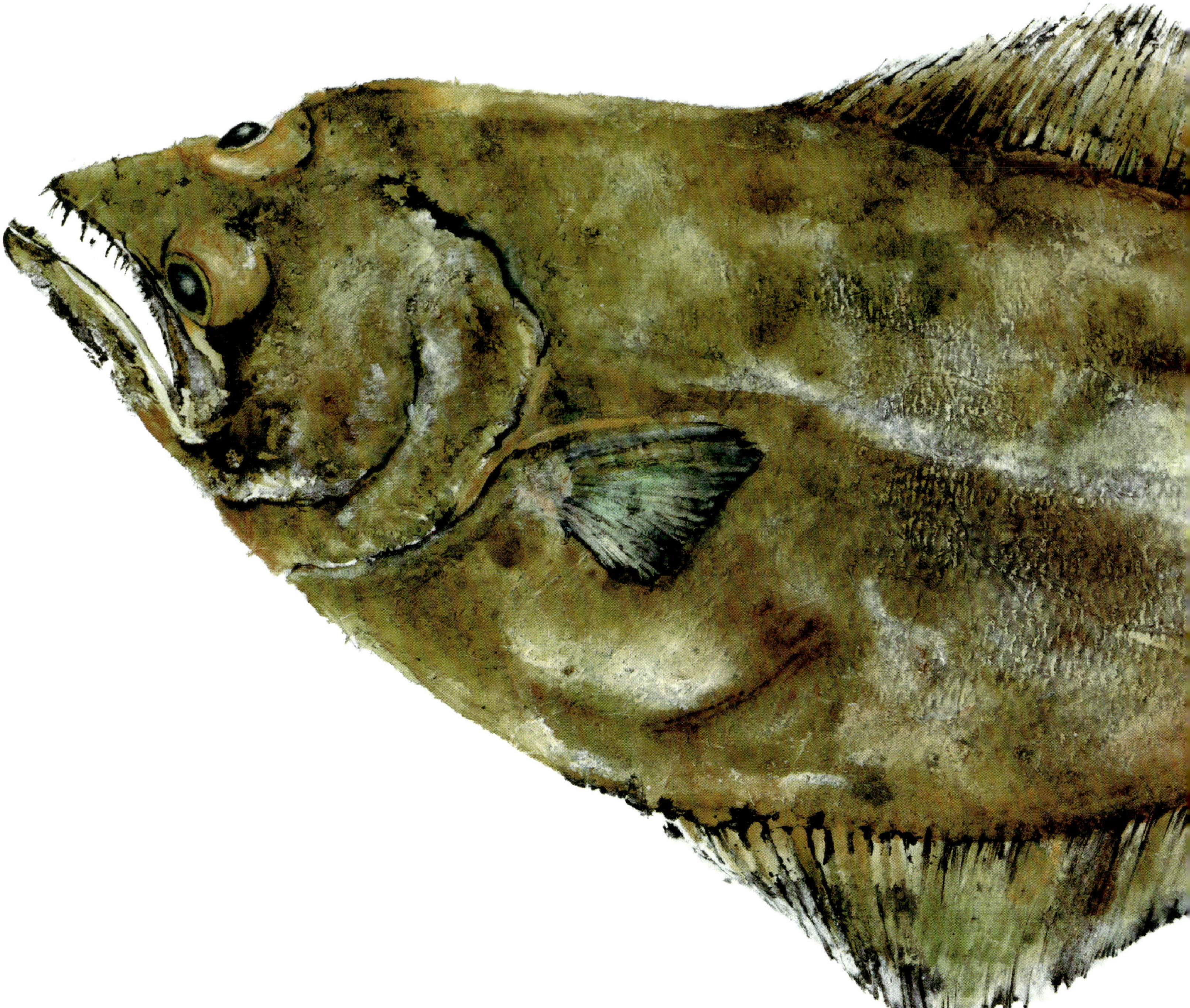

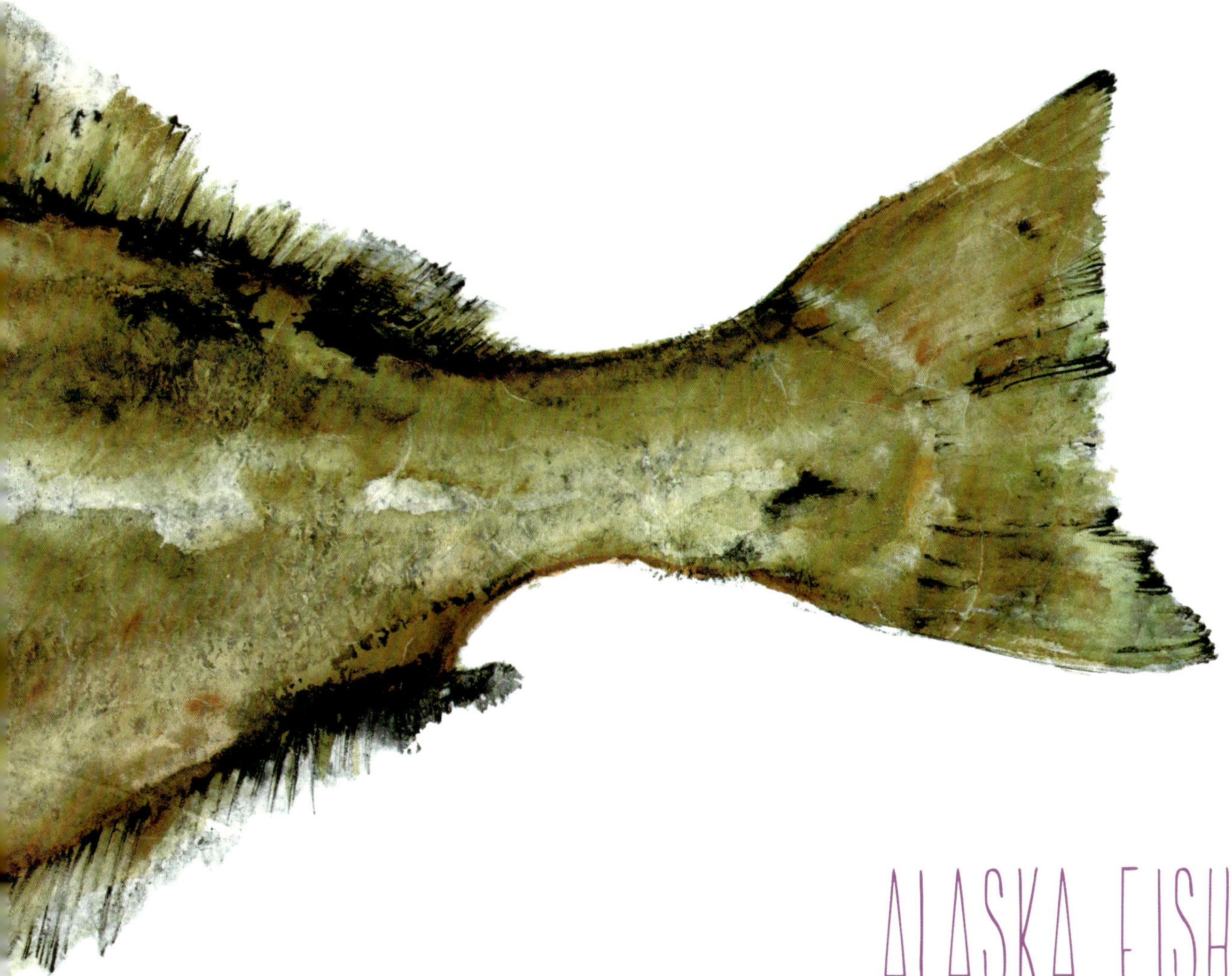

ALASKA FISH GYOTAKU

CHINOOK SALMON

(*Oncorhynchus tshawytscha*)

COMMON NAMES: Blackmouth, king, quinnat, tule, and tyee

LIFESPAN: Three to seven years

WEIGHT: Up to 30 pounds

HABITAT: Streams and rivers; open ocean

At sea, adults are distinguished by the irregular black spots on the back, the dorsal fins, and both lobes of the caudal or tail fin. Chinook salmon, commonly called king salmon, also have a black pigment along the gum line, hence the name "blackmouth." When spawning, king salmon turn red, to copper, to almost black.

The Chinook salmon is the largest of all Pacific salmon. On average, the Chinook salmon weighs up to or exceeding 30 pounds and measures close to 36 inches.

Salmon begin their lives in streams and rivers. After a period of one to three years, the small salmon, known as smolt, move to the ocean where they spend their adult lives. After three to seven years, the female chinook returns to the place where she was born and lays her eggs in a gravel bed nest called a redd. This is where the male king salmon will fertilize the eggs. The average lifespan of the king salmon is from three to seven years.

The chinook is the official fish of Alaska. The species name comes from the common name used among Native groups in Alaska and Siberia. Its common names are blackmouth, king, quinnat, tule, and tyee.

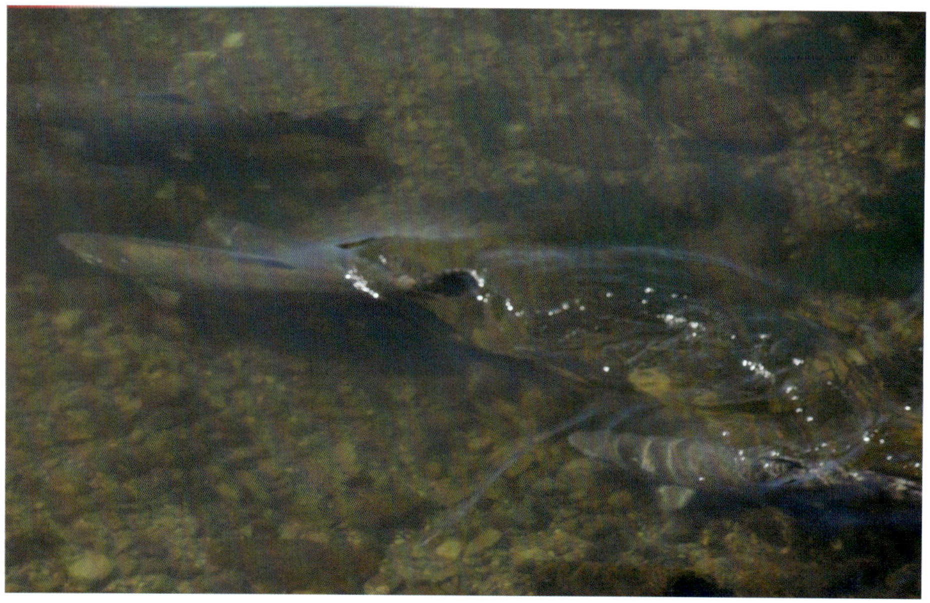

Salmon swimming under water, Herring Cove, Ketchikan, Alaska.

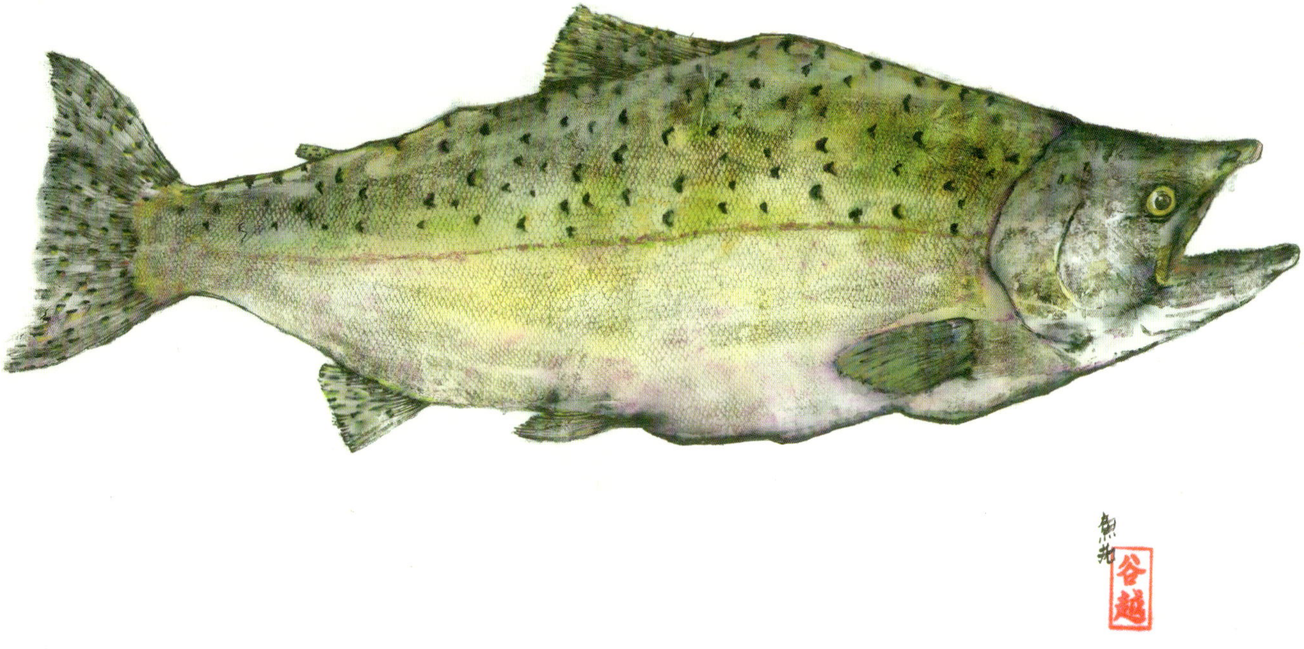

Chinook salmon with watercolor.

COHO SALMON

(Oncorhynchus kisutch)

COMMON NAME: Silver salmon

LIFESPAN: Up to four years

WEIGHT: 8 to 30 pounds

HABITAT: Streams and rivers; open ocean

The coho salmon is a beautiful fish with bright silver coloring when in the ocean. It has spots on the upper lobe of its tail, as well as on its back. Its mouth is dark with a white gum line. On average, it weighs about 8 pounds, but can reach up to 30 pounds.

This anadromous fish spends the first one to three years of its life in the ocean, where it boasts a silver color and dark-blue back. When the coho transitions to its spawning phase, it swims inland to the streams. In freshwater, the silver salmon changes color. Its silver sides turn dark red, and its head and back become bluish green. During the spawning phase, the male's back slightly arches and the female's belly turns a rosy pink color.

Both sexes acquire a hooked jaw called a kype. The average age for male coho salmon to spawn is at three years. Some males spawn at two years of age. They are known as jack silvers. The coho can live up to four years.

The coho salmon is a symbol of life and sustenance for many Native Alaskan groups. Other names for the coho are silver salmon, or silvers.

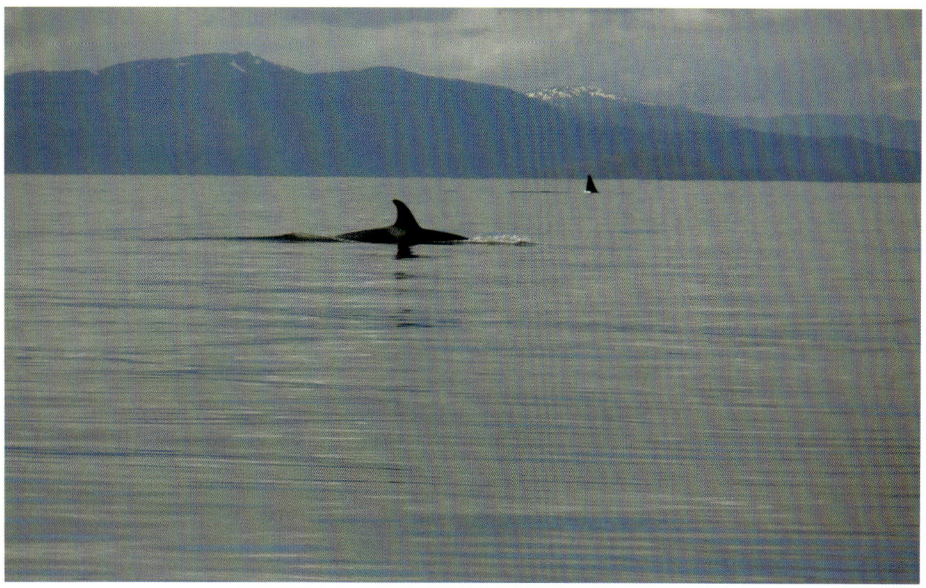

Orcas swimming near Prince of Wales Island, Alaska.

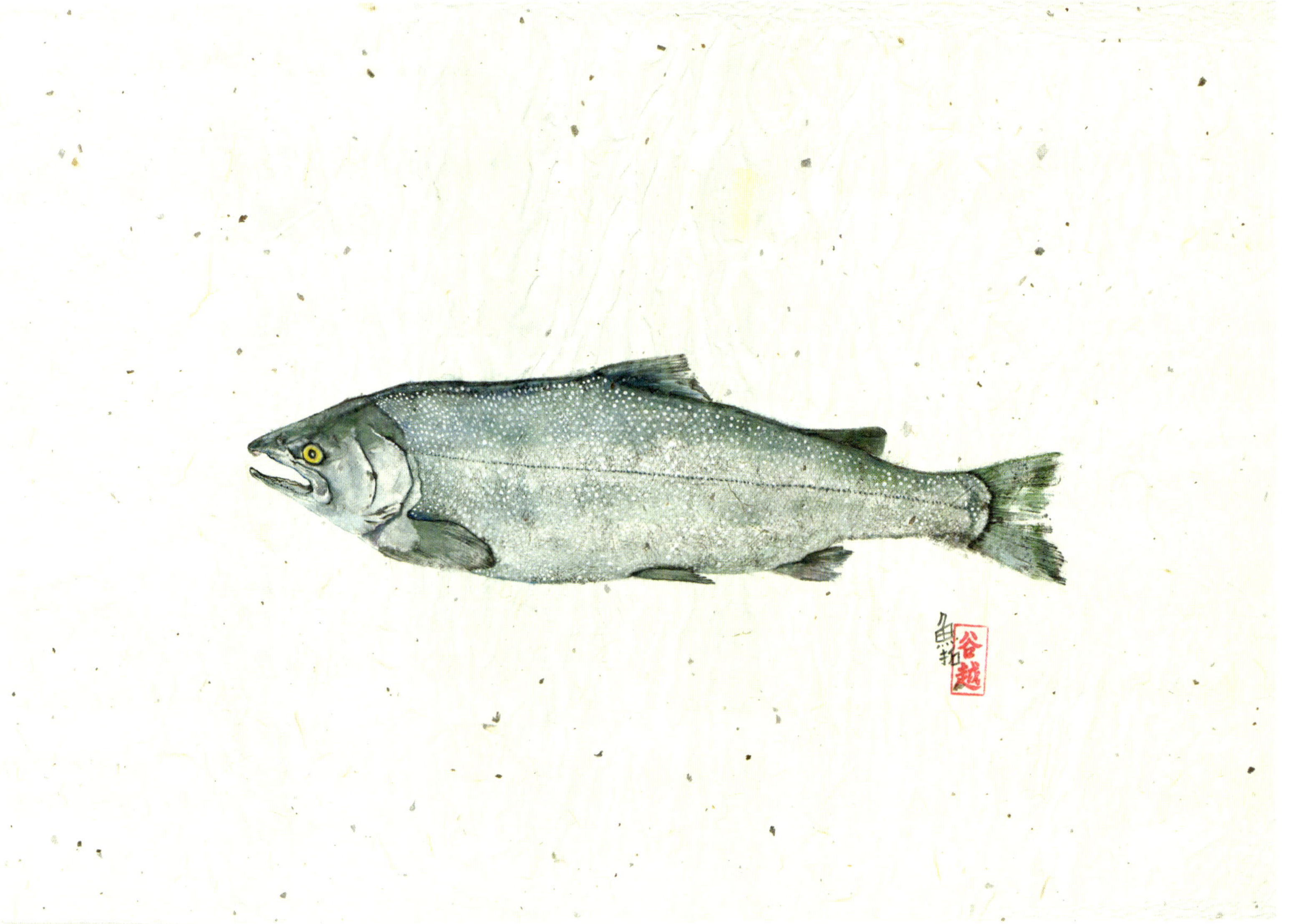

Coho salmon.

HUMPBACK SALMON

(*Oncorhynchus gorbuscha*)

COMMON NAME: Pink salmon

LIFESPAN: Two years

WEIGHT: 3.5 to 5 pounds

HABITAT: Streams and rivers; open ocean

The humpback salmon is characterized by large oval-shaped black spots on its back and a V-shaped tail. The mouth of the humpback salmon is white with black gums. Its scales are tiny compared to other Pacific salmons'. The humpback salmon has an average weight of 3.5 to 5 pounds. It is the shortest of all Pacific salmon and averages only 20 to 25 inches in length.

It has a strict two-year cycle that prevents the odd/even cycles from interbreeding. The humpback salmon is a beautiful bright-silver color when in the ocean and a dull green to yellowish green when it is spawning in the rivers. The back of the humpback (pink) salmon becomes very hunched and turns a dark green during spawning season (late June to mid-October).

The humpback salmon has a range extending from northern California up through Alaska and reaching over to the Pacific Rim of Asia.

In Russian, the word *gorbúša* (*горбуша*) means "humpy." Native Alaskans call the humpback salmon the humpy, or a pink because of its pink meat. Typically, the pink salmon is canned and marketed as such.

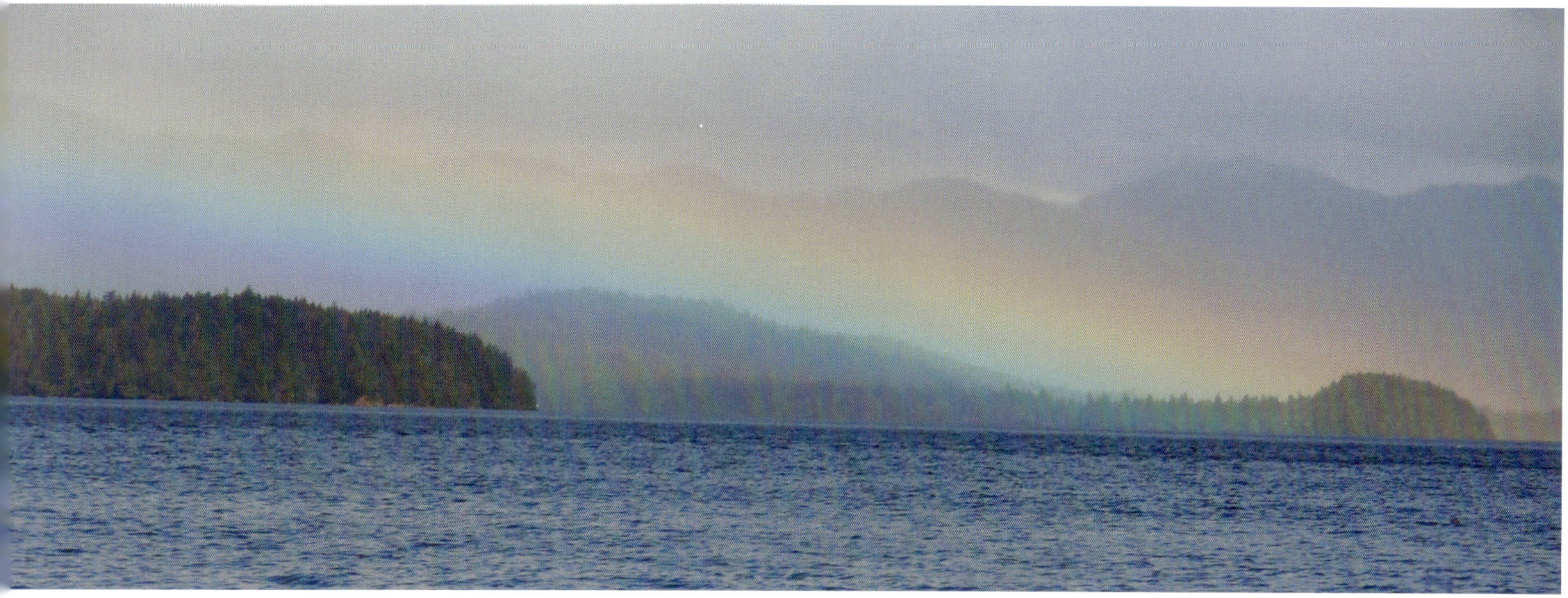

Rainbow over Grant Island, Alaska.

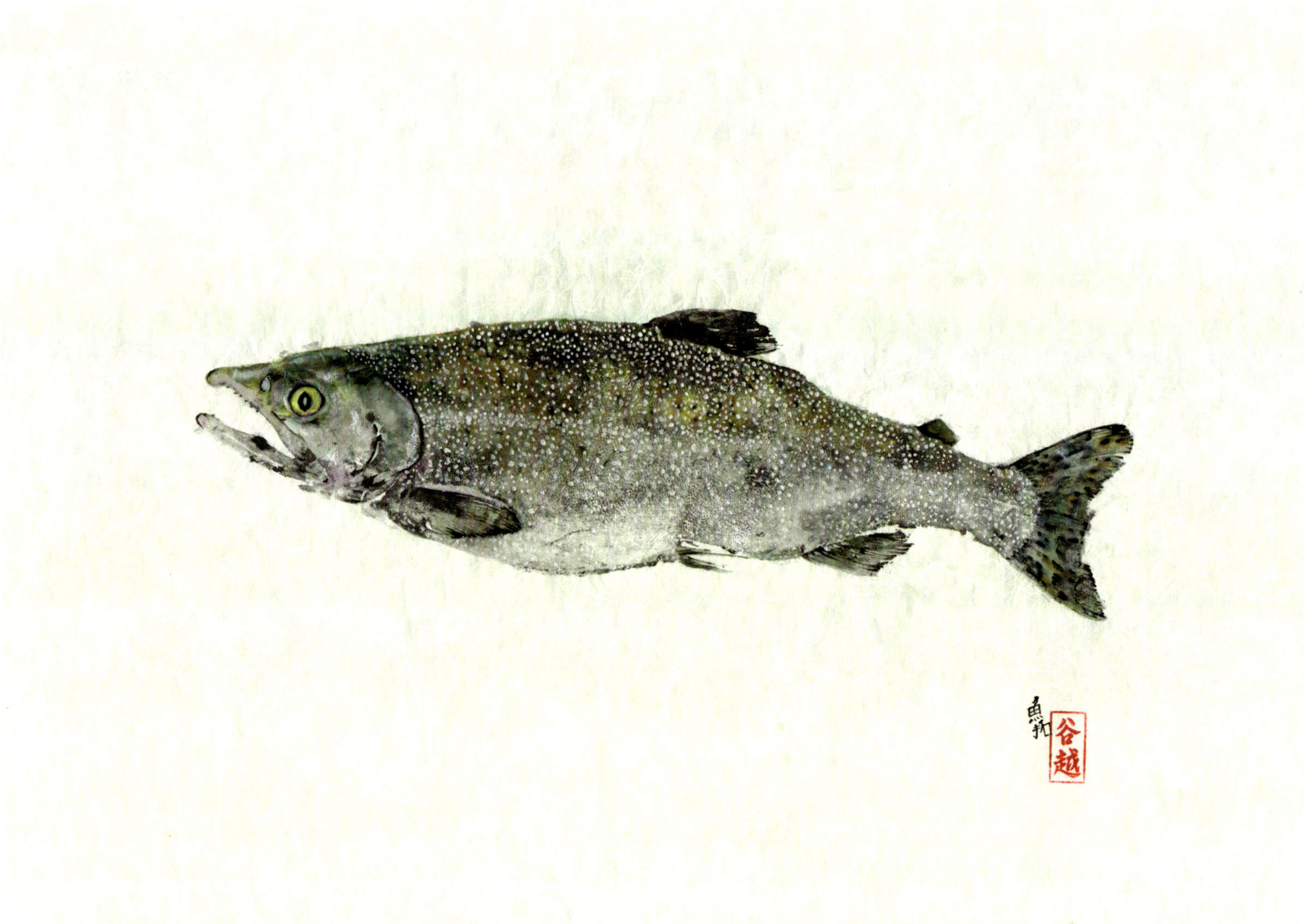

Humpback salmon.

Black Rockfish

(*Sebastes melanops*)

COMMON NAMES: Black snapper, black rock cod, and black bass

LIFESPAN: Up to forty-nine years

WEIGHT: 5 to 11 pounds

HABITAT: Rocky structures from 150 to 366 meters deep

The black rockfish, or black bass, is one of thirty-two species of rockfish in the family *Scorpaenidae* found off the coast of Alaska. As a member of the *Scorpaenidae* family, its spines are venomous, although less so than the spines of the quillback rockfish. It can grow up to 25 inches long and averages 5 pounds in weight. It can weigh up to 11 pounds.

The black rockfish is a pelagic fish. It is a schooling species that inhabits rocky structures and areas full of structure and kelp. The term *pelagic* comes from the Greek word *pélagos*. This word means "open sea" and describes fish that reside in the water column from the surface of the water to just above the ocean's floor. Pelagic fish generally live from seven to thirty years.

The black rockfish's range is from the Aleutian Islands in Alaska down to southern California. It is typically found at depths around 150 meters, but can be much deeper (up to 366 meters).

Interestingly, the black rockfish is known to live up to forty-nine years of age, but only becomes sexually mature between six to eight years of age. The black rockfish, like all other rockfish, gives birth to live young.

In Greek, its scientific name means "magnificent" (*sebastos*) "black" (*melas*) "face" (*ops*). In Southeast Alaska, the Tlingit have indigenous names for the black rockfish. They call it *lit.isdúk*. The Alutiiq call it the *tu-kuq*.

Common names include the black snapper, black rock cod, and black bass.

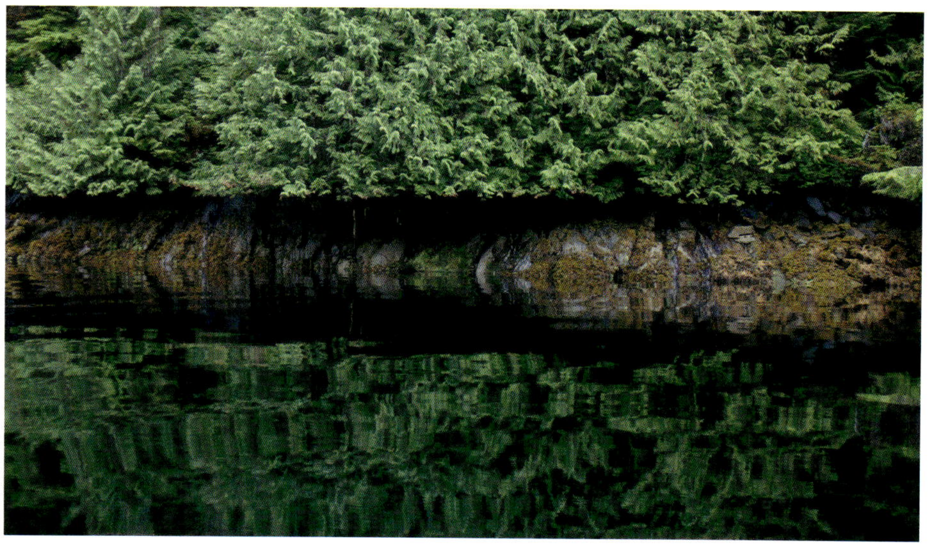

Cliff reflection.

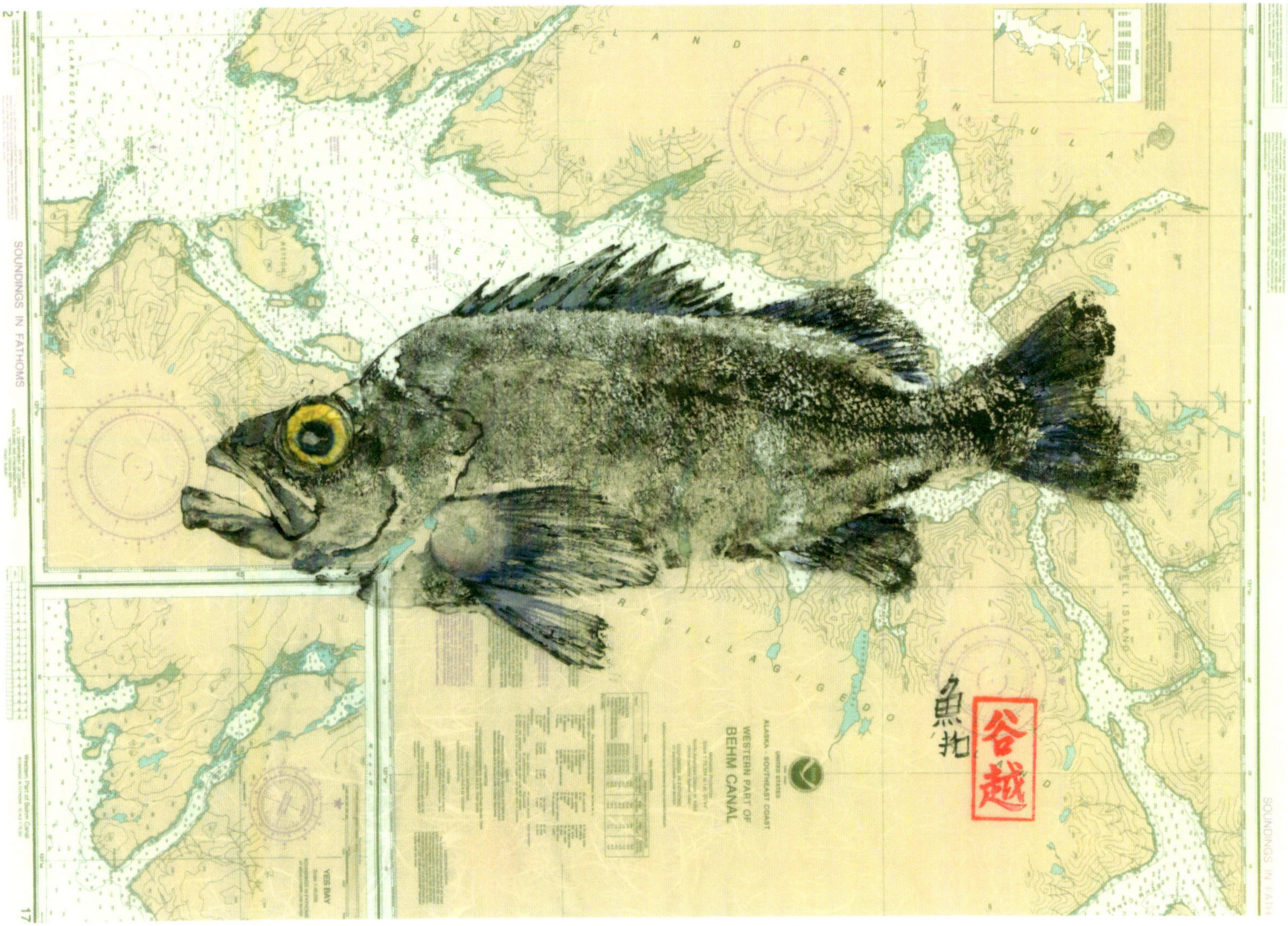

Black rockfish on nautical chart.

CANARY ROCKFISH

(Sebastes pinniger)

COMMON NAMES: Fantail, orange rockfish, and red snapper

LIFESPAN: Up to eighty-four years

WEIGHT: Up to 10 pounds

HABITAT: Rocky habitat from 80 to 200 meters deep

The canary rockfish is a stunning orange-yellow rockfish with three distinctive stripes that seem to flow from its eyes to its body. The adult canary rockfish is typically around 24 inches long and about 7 pounds in weight. It has been known to weigh up to 10 pounds and measure up to two and a half feet in length.

The canary rockfish lives in groups around rocks or pinnacles in water from 80 to 200 meters deep. Juvenile canary rockfish can live in open waters for months and allow the current to move them from place to place. Adult canary rockfish live in rocky habitat from 80 to 200 meters deep. The canary rockfish can live up to a surprising eighty-four years!

Unlike some rockfish that tend to stay in one area, the canary rockfish has been tagged and found to travel up to 275 miles in four years. The canary rockfish has a range from Cape Colnett in Baja California up to Cape San Bartolome in Alaska.

Because these white-fleshed fish are targeted by commercial and recreational anglers, and sometimes caught incidentally by halibut trawlers, the canary rockfish has suffered a population decrease. Additionally, some of the rockfish, such as the canary and the yelloweye, do not reach maturity until they are twenty years old and overfishing drastically affects their populations. In fact, canary rockfish are a threatened species in Puget Sound and George Basin in Washington. In Southeast Alaska, bag limits of retaining canary rockfish have been limited to one fish daily.

Common names of the canary rockfish are fantail, orange rockfish, and red snapper.

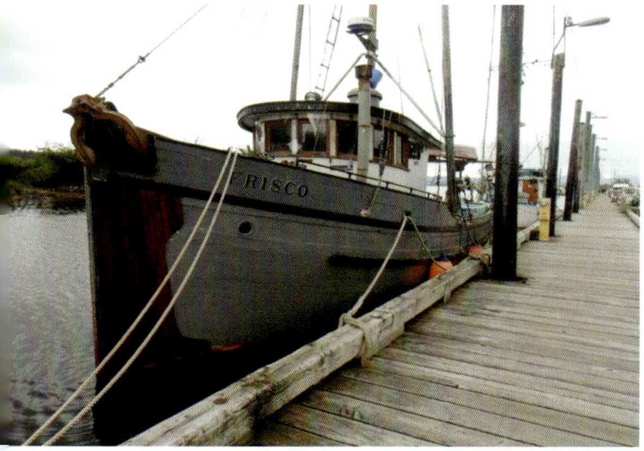

Wooden trawler on Prince of Wales Island, Alaska.

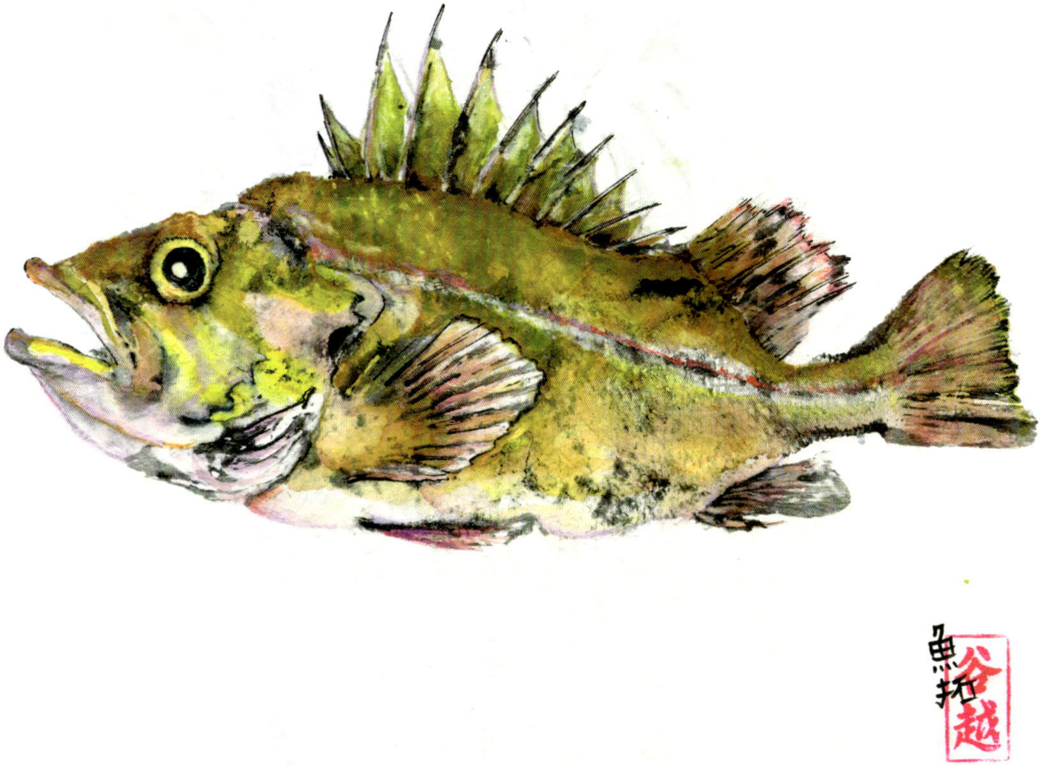

Canary rockfish.

COPPER ROCKFISH

(*Sebastes caurinus*)

COMMON NAMES: Chucklehead, fighting bob, never-dies, white belly, and white gopher

LIFESPAN: Up to fifty-five years

WEIGHT: Up to 6 pounds

HABITAT: Rocky structure from 120 to 183 meters deep

The copper rockfish is a sturdy fish with long dorsal spines. Its body is dark brown to olive in color with a pinkish and yellowish hue. If it is exposed to the air, it becomes more and more coppery and pink. According to the Alaska Fisheries Science Center, copper rockfish are commonly olive, dark brown, or copper pink in color. In California, the copper rockfish is typically bright red. Some northern areas have copper rockfish that are almost all black in color. These variations are interesting to note. The distinguishing factor is that the rear two-thirds of its lateral line includes a light-color strip that differentiates it from the gopher and quillback rockfish. There are also several bars of copper orange to yellow colors extending back from its eyes.

The copper rockfish can grow to be about 22 inches long and weigh around 6 pounds. It can live to be fifty-five years old.

The copper rockfish inhabits waters from the Gulf of Alaska to Baja California. The copper rockfish is typically found in rocky structure at depths of 120 meters to 183 meters.

The male copper rockfish matures between four to seven years. The female matures between four to eight years. Like all rockfish, she gives birth to live young. Its diet consists of planktonic crustaceans, shrimp, crabs, squid, and octopus.

Interestingly, the copper rockfish stays within one mile of its chosen home for its entire life.

Common names for the copper rockfish are the chucklehead, fighting bob, never-dies, white belly, and white gopher.

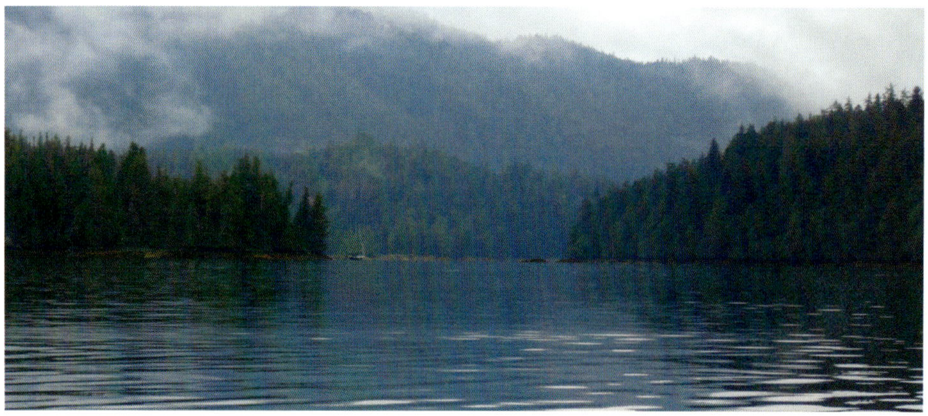

Moser Bay, Alaska.

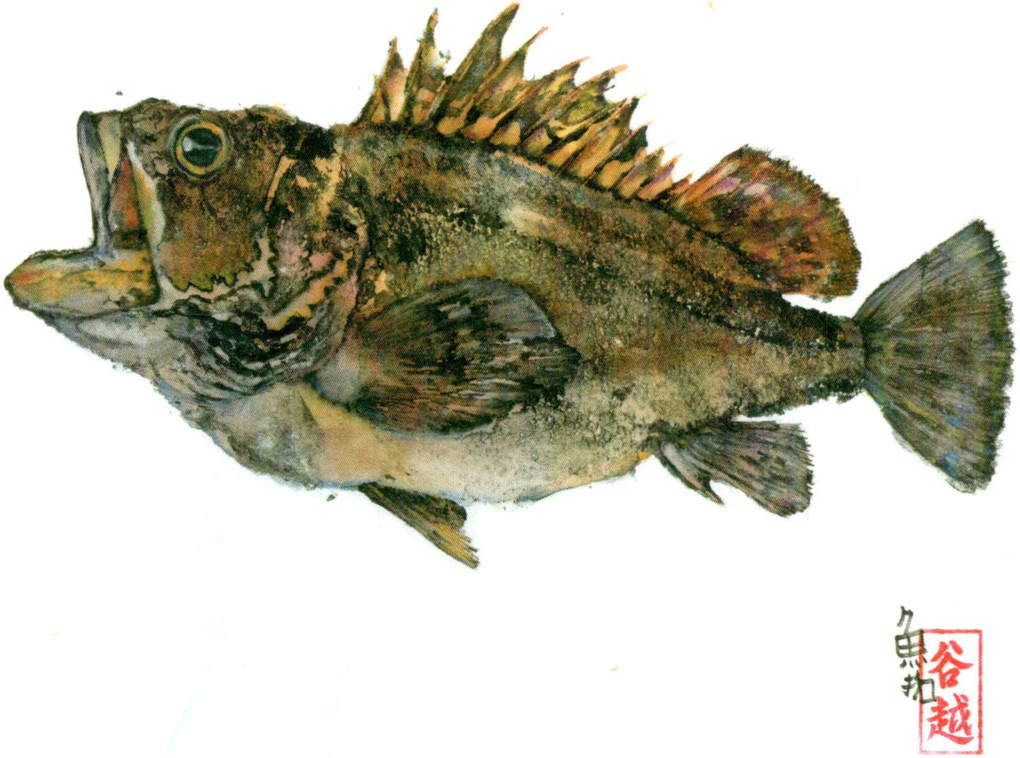

Copper rockfish.

DUSKY ROCKFISH

(*Sebastes variabilis*)

COMMON NAMES: None

LIFESPAN: Up to fifty years

WEIGHT: 3 to 4 pounds

HABITAT: Deep offshore banks from 100 to 300 meters deep

The dusky rockfish is a member of the *Scorpaenidae* family that gets its name from its members' scorpion-like fin spines. This schooling rockfish is brownish blue on most of its body with a whitish, pink-orange belly. It has two dark bars on each of its cheeks. Typically, it is a smaller rockfish, with its longest length at about 20 inches. It weighs between 3 and 4 pounds.

The dusky rockfish lives in deep offshore banks from 100 to 300 meters deep. According to the Alaska Fisheries Science Center, the dusky rockfish can live up to fifty years.

The dusky rockfish has a wide habitat range. It can be found from the coast of central Oregon to the Aleutian Islands in Alaska and all the way over to the east coast of Kamchatka in Russia.

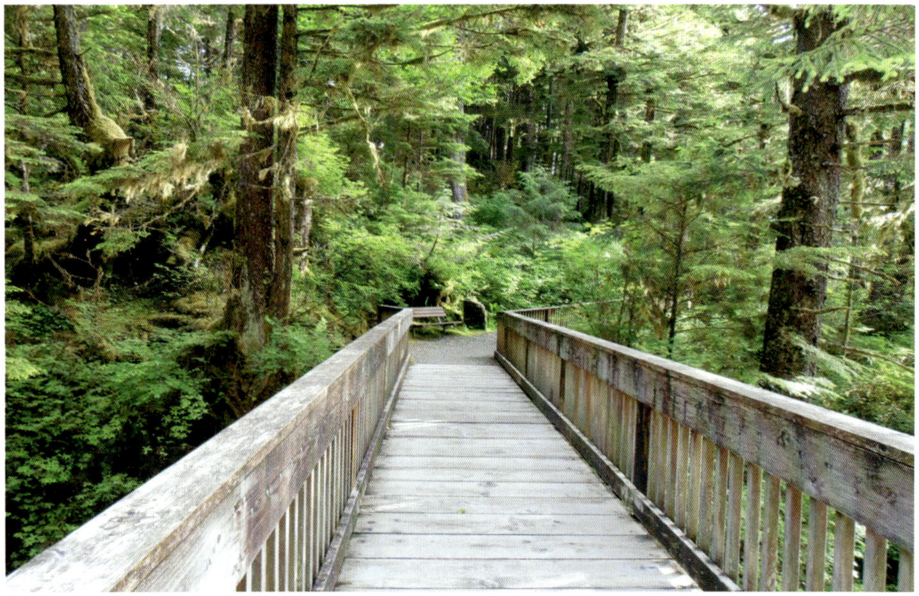

Bridge at Settler's Cove, Ketchikan, Alaska.

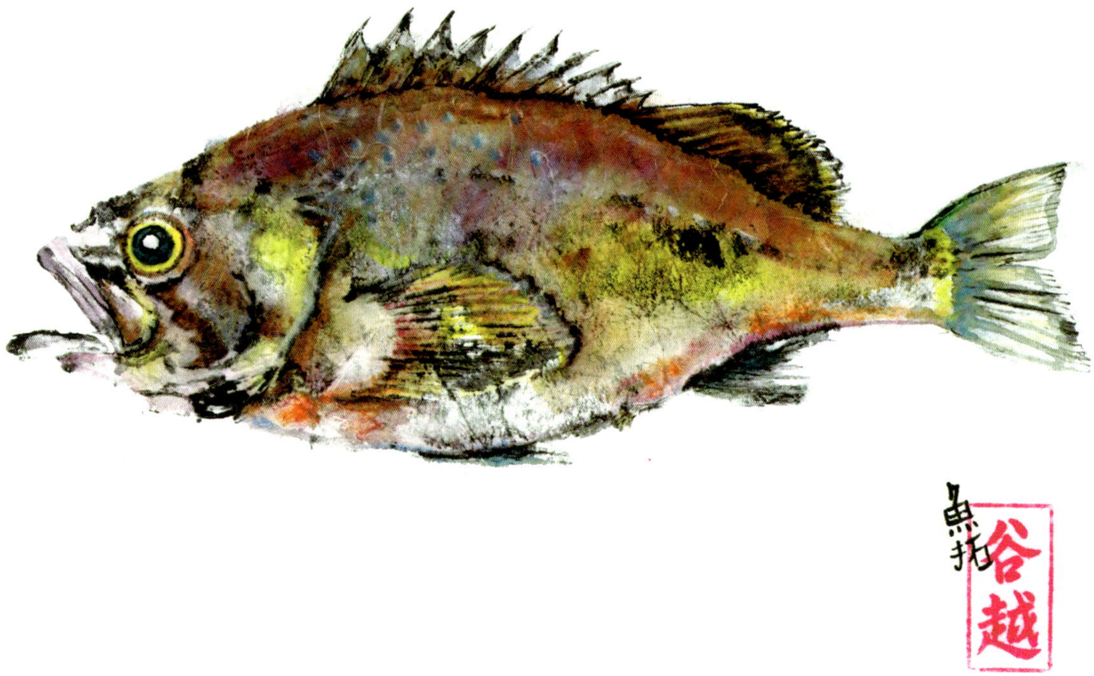

Dusky rockfish.

PACIFIC OCEAN PERCH

(*Sebastes alutus*)

COMMON NAMES: Rose perch, Pacific rockfish, red bream, or red perch

LIFESPAN: Up to 114 years

WEIGHT: Up to 4.5 pounds

HABITAT: Juveniles—shallow water to 36 meters deep; adults—91 to 822 meters deep

In Greek, *alutus* means "unwashed" or "speckled," which is very appropriate for this fish. The Pacific Ocean perch is a reddish fish with deep-red, olive-green, brown, or black mottling, becoming more pronounced as the fish matures. The juvenile Pacific Ocean perch is a light-red color with a whitish belly. This perch can weigh up to four and a half pounds and measure up to 21.2 inches.

The adult Pacific Ocean perch generally are found at depths of 91 to 822 meters, while the juveniles are found in shallower waters around 36 meters deep. Its range is from northern California to the Aleutian Islands in Alaska and over to the Gulf of Alaska. They have been found in Honshu, Japan.

These are long-lived rockfish that can live to be 114 years old, according to the Alaska Fisheries Science Center.

The Pacific Ocean perch is commonly marketed as a whole fish, used for surimi, and as skinless, boneless fillets.

The Pacific Ocean perch is also called the rose perch, the Pacific rockfish, red bream, or red perch.

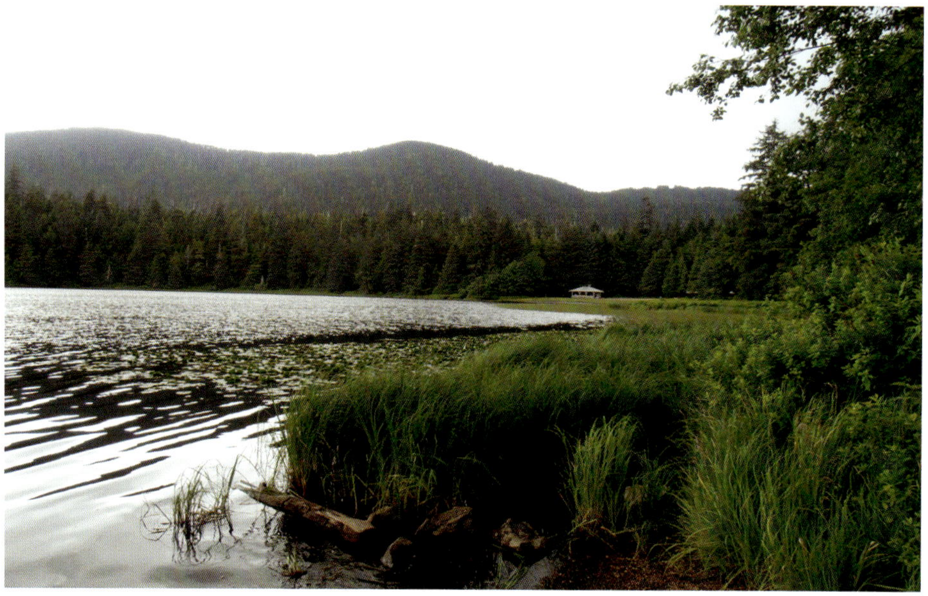

Ward Lake, Ketchikan, Alaska.

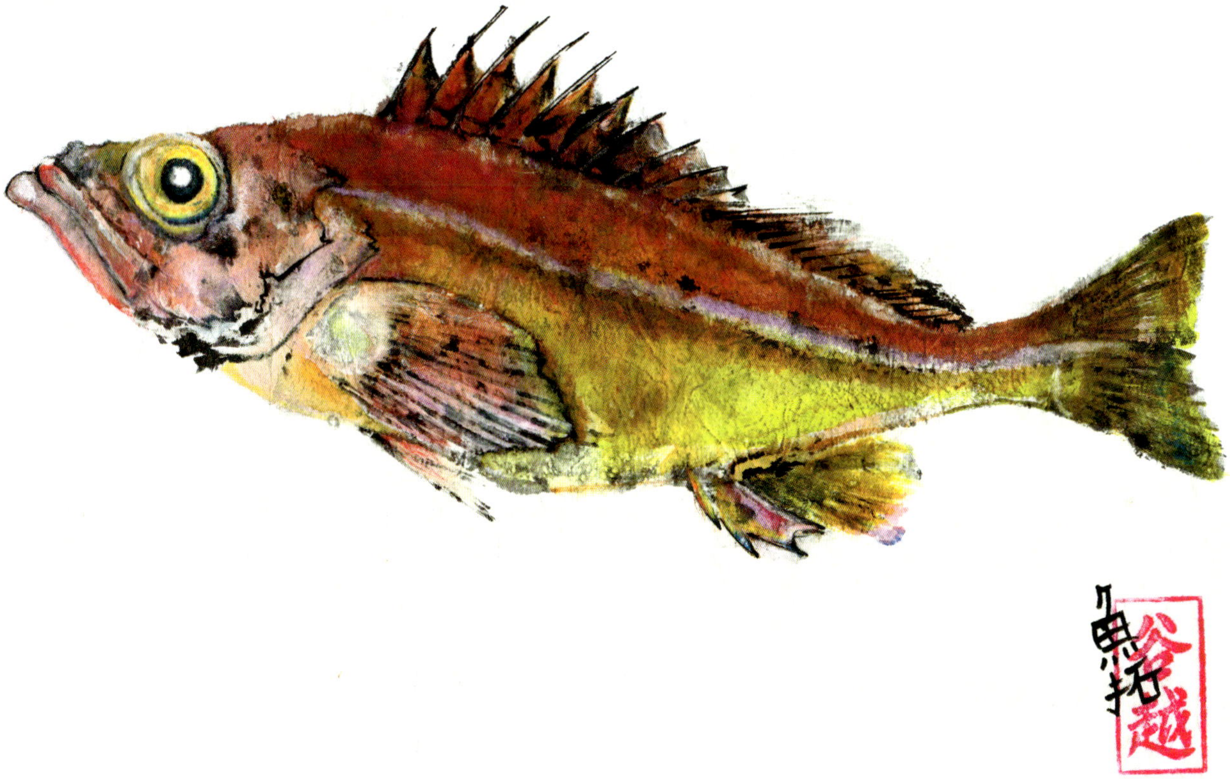

Pacific Ocean perch.

QUILLBACK

(Sebastes maliger)

COMMON NAMES: Orange-speckled rockfish, speckled rockfish, spiny back, and yellow-backed rockfish

LIFESPAN: Up to ninety years

WEIGHT: 2 to 7 pounds

HABITAT: Rocky structures from 75 to 272 meters deep

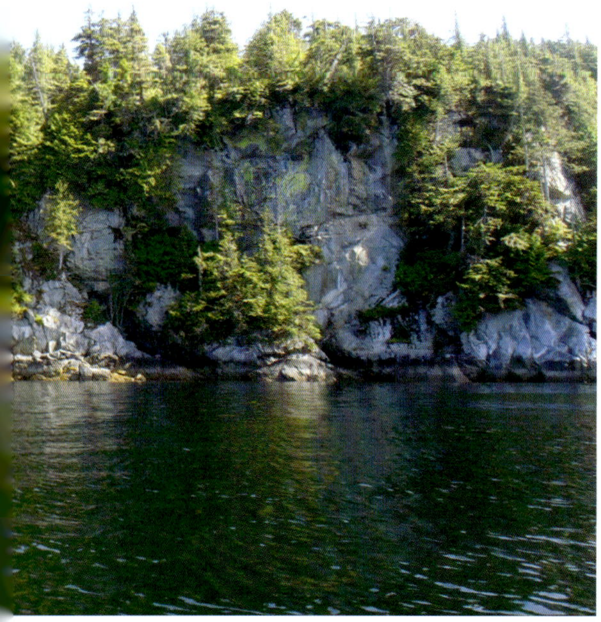

Tatoosh Rocks, Ketchikan, Alaska.

The quillback is a mottled brown, orange, and yellow color. This fish generally weighs between 2 to 7 pounds and grows to around 24 inches in length. The quillback has a lifespan of up to ninety years!

The quillback rockfish lives in rocky structures from 75 to 272 meters deep. Its range runs from the Channel Islands in southern California up to the Gulf of Alaska.

The main diet of the quillback consists of crustaceans and herring. The quillback is not overly territorial, nor does it like to live in groups.

The quillback rockfish is named for its long, sharp spines on its dorsal fin. At the base of its dorsal spines, the quillback has bilateral venom glands that secrete poison into the spines as a defense against predators. Each dorsal spine contains a glandular groove through which crystalline molecules of venom are released when the quillback feels threatened.

The tips of the spines are very pointed and sharp. If a person is stung, it is very painful. Although life-threatening reactions are highly uncommon, being stung by a quillback rockfish should be taken seriously. If one is stung, there can be swelling, throbbing, burning, and fever. Care must be taken to remove any part of the spine that breaks off into the wound, or subsequent infection will occur.

A person might have an allergic reaction to the quillback venom that might result in hives, swelling, rashes, and itching. In severe cases, this could result in breathing problems, tightness and/or pain in the chest, and swelling of the lips, mouth, or throat. In this case, immediate medical attention is necessary.

Common names for the quillback are orange-speckled rockfish, speckled rockfish, spiny back, and yellow-backed rockfish.

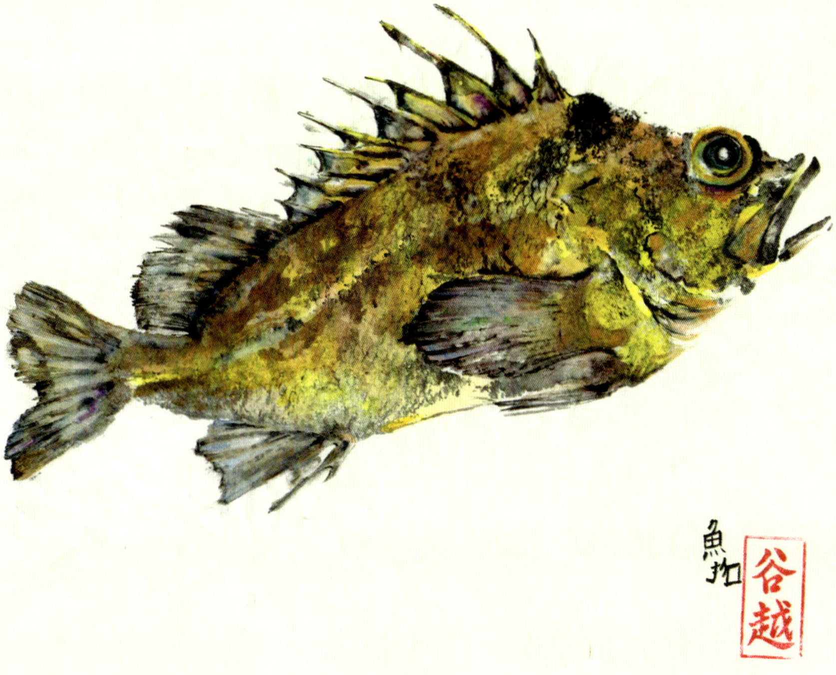

Quillback rockfish.

Silvergray Rockfish

(*Sebastes brevispinis*)

COMMON NAMES: Greenie, grouper, longjaw, rock salmon, silverbelly, silverside, and snapper

LIFESPAN: Up to eighty-two years

WEIGHT: Up to 11 pounds

HABITAT: Rocky structures from 17 to 300 meters deep

The silvergray rockfish is gray with red, pink, or brown-pink coloration, and frequently has small dark spots. Its body is grayish in color, while its belly is white with shades of orange or pink. The silvergray is commonly known as the longjaw rockfish and is characterized by a long lower jaw that extends far beyond the upper jaw. Its Latin name, *brevispinis* means "short spine."

According to the Alaska Fisheries Science Center, this large rockfish can grow up to 29.2 inches in length. It can weigh up to 11 pounds.

The silvergray rockfish lives in saltwater in rocky habitat from 17 to 300 meters deep. Its range is from the western Gulf of Alaska south through Oregon and down to central Baja California. It has been known to live up to eighty-two years.

Common names for the silvergray rockfish include greenie, grouper, longjaw, rock salmon, silverbelly, silverside, and snapper.

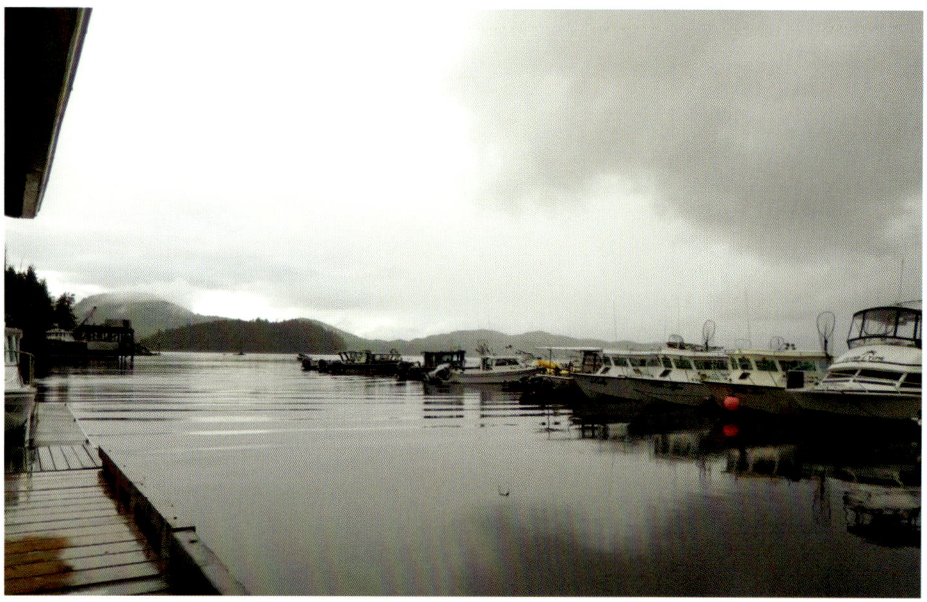

Charter boats at Knudson Cove, Ketchikan, Alaska.

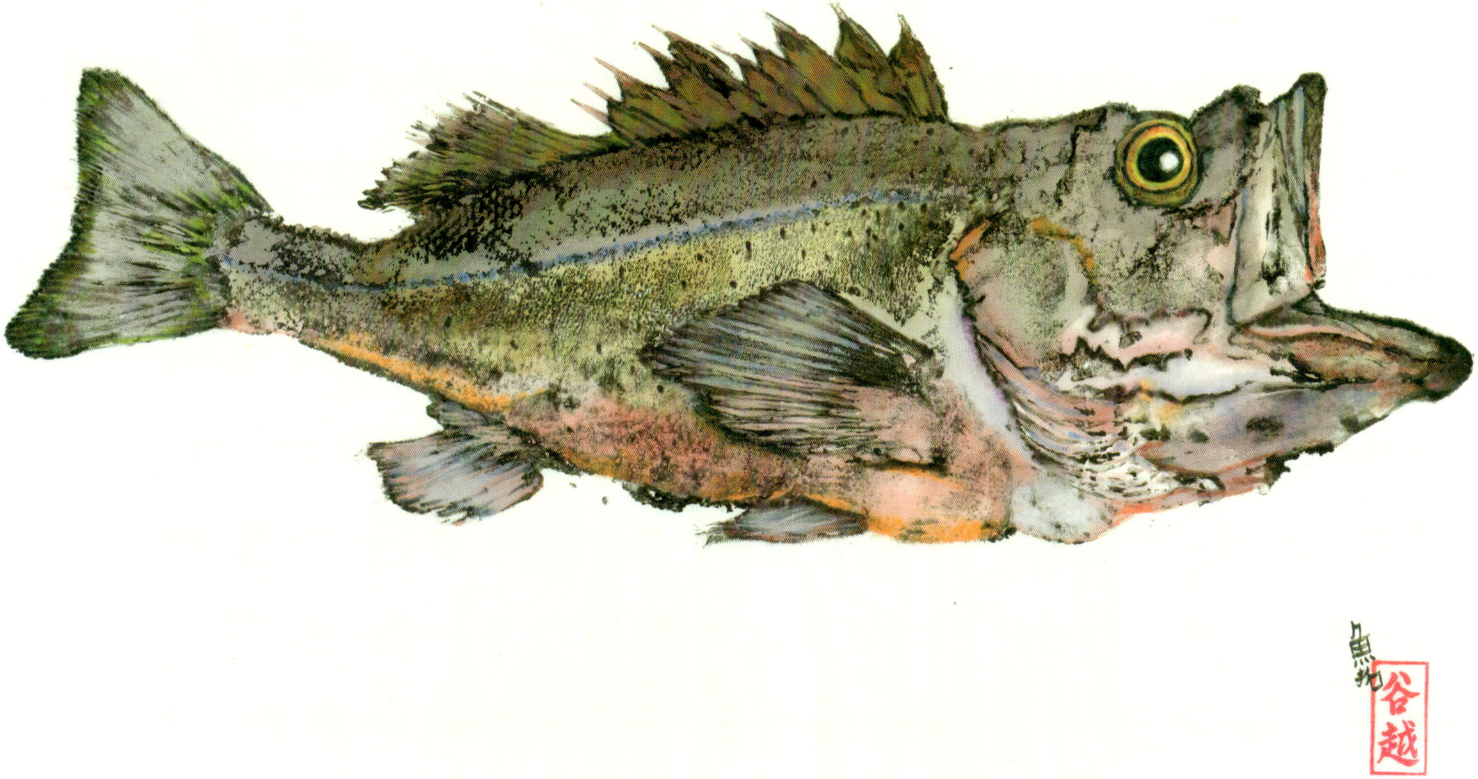

Silvergray rockfish.

TIGER ROCKFISH

(*Sebastes nigrocinctus*)

COMMON NAMES: Black-banded rockfish, banded rockfish

LIFESPAN: Up to 116 years

WEIGHT: Up to 5 pounds

HABITAT: Rocky structures, underground crevices, and caves 18 to 298 meters deep

The tiger rockfish, named for the five dark-red stripes running vertically along both sides of its body, is a beautiful member of the *Scorpaenidae* family. Its body is a light-pink color and has a line of tall, strong spines along its back. The head of the tiger rockfish boasts bony ridges.

It can weigh up to 5 pounds and measure up to 2 feet in length.

The tiger rockfish lives in rocky structures, underground crevices, and caves 18 to 298 meters down below the surface of the ocean. The tiger rockfish is generally solitary. Its diet consists of shrimp, herring, juvenile rockfish, and rock crabs. In order to eat, the tiger rockfish waits for water currents to bring food near its territory.

The black-banded rockfish has a range from Prince William Sound, Alaska, to the waters off Kodiak Island and down south to Point Buchon, central California.

The tiger rockfish has been found to live up to 116 years!

Common names for the tiger rockfish are the black-banded rockfish and the banded rockfish.

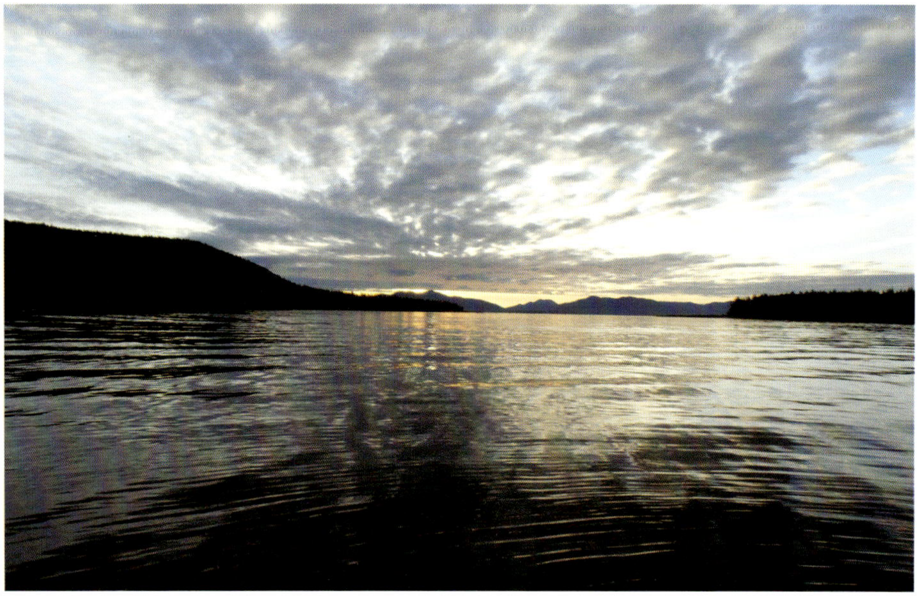

Sunset at Grant Island, Alaska.

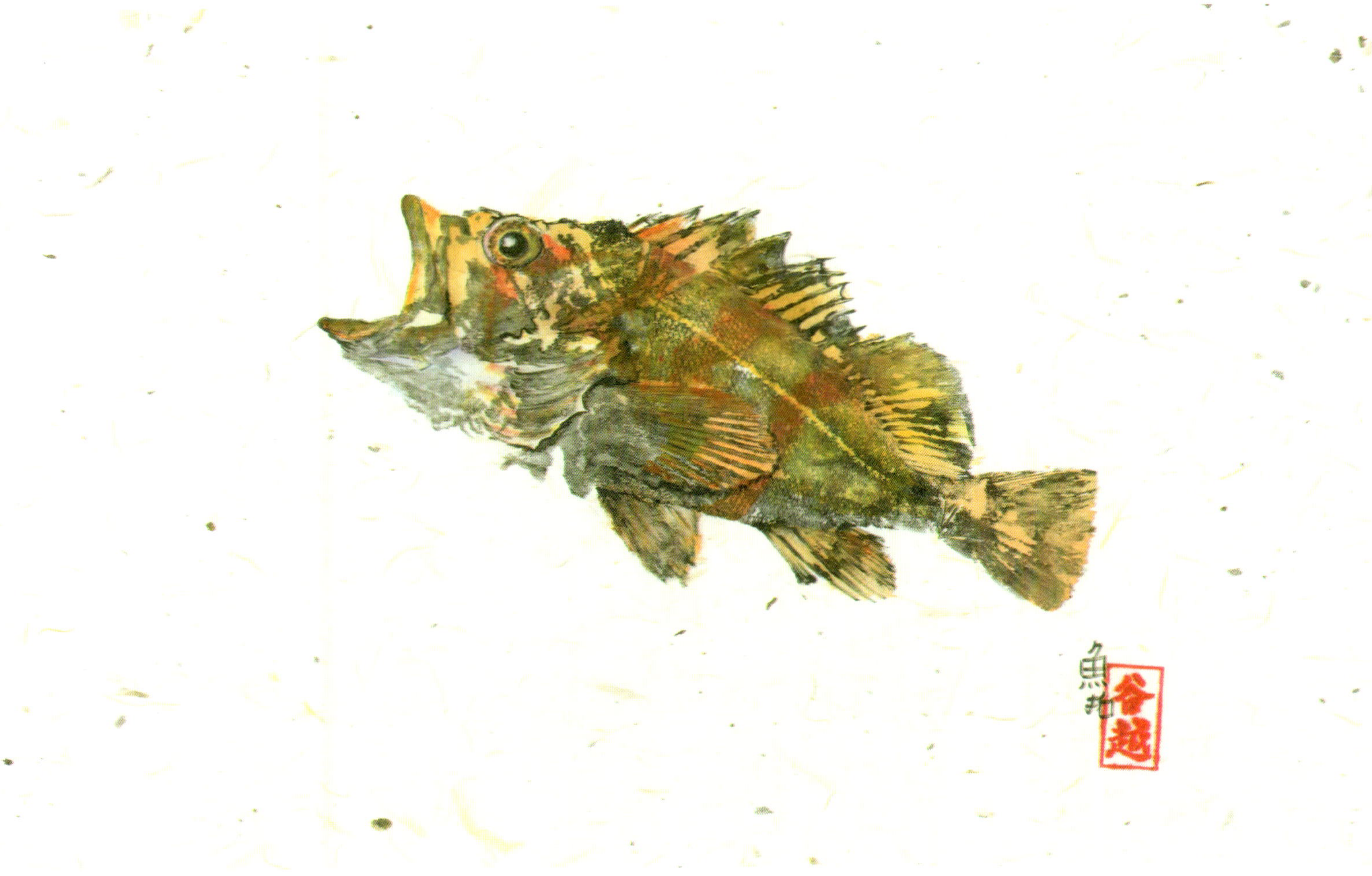

Tiger rockfish.

YELLOWEYE ROCKFISH

(*Sebastes ruberrimus*)

COMMON NAMES: Red rock cod, Pacific red snapper, and yellow belly

LIFESPAN: Up to 121 years

WEIGHT: Up to 40 pounds

HABITAT: Rocky structures from 25 to 475 meters deep

Yelloweye rockfish are orange-red to orange-yellow in color and may have black on their fin tips. Their eyes are bright yellow. Adults usually have a light to white stripe on the lateral line; juveniles have two light stripes, one on the lateral line and a shorter one below the lateral line. Approximately half of adult yelloweye rockfish reach maturity. The juveniles look very different from the adults and were once thought to be two different species.

In Southeast Alaska, the Tlingit have indigenous names for the yelloweye. They call the yelloweye *léik'w*. The Alutiiq call the yelloweye *ushmaq*.

The yelloweye rockfish can reach up to three and a half feet in length and about 40 pounds in weight.

The yelloweye rockfish lives in rocky areas from 25 to 475 meters deep. It is a very long-lived fish. It has been found to live up to 121 years, according to the National Oceanic and Atmospheric Administration.

The yelloweye rockfish has a range from the Aleutian Islands in Alaska on south to northern Baja California.

According to the California Department of Fish and Wildlife, the yelloweye rockfish was declared overfished in 2002 and has since been managed under a federal rebuilding plan to try to allow its population to recover. In Washington, the Department of Fish and Game prohibits retention of the yelloweye rockfish. In Alaska, there is a strict limit of one yelloweye per angler per year. Because the yelloweye reaches sexual maturity between ten to twenty years of age, it could take one hundred years to make a significant recovery of its population.

Common names for the yelloweye rockfish are red rock cod, Pacific red snapper, and yellow belly.

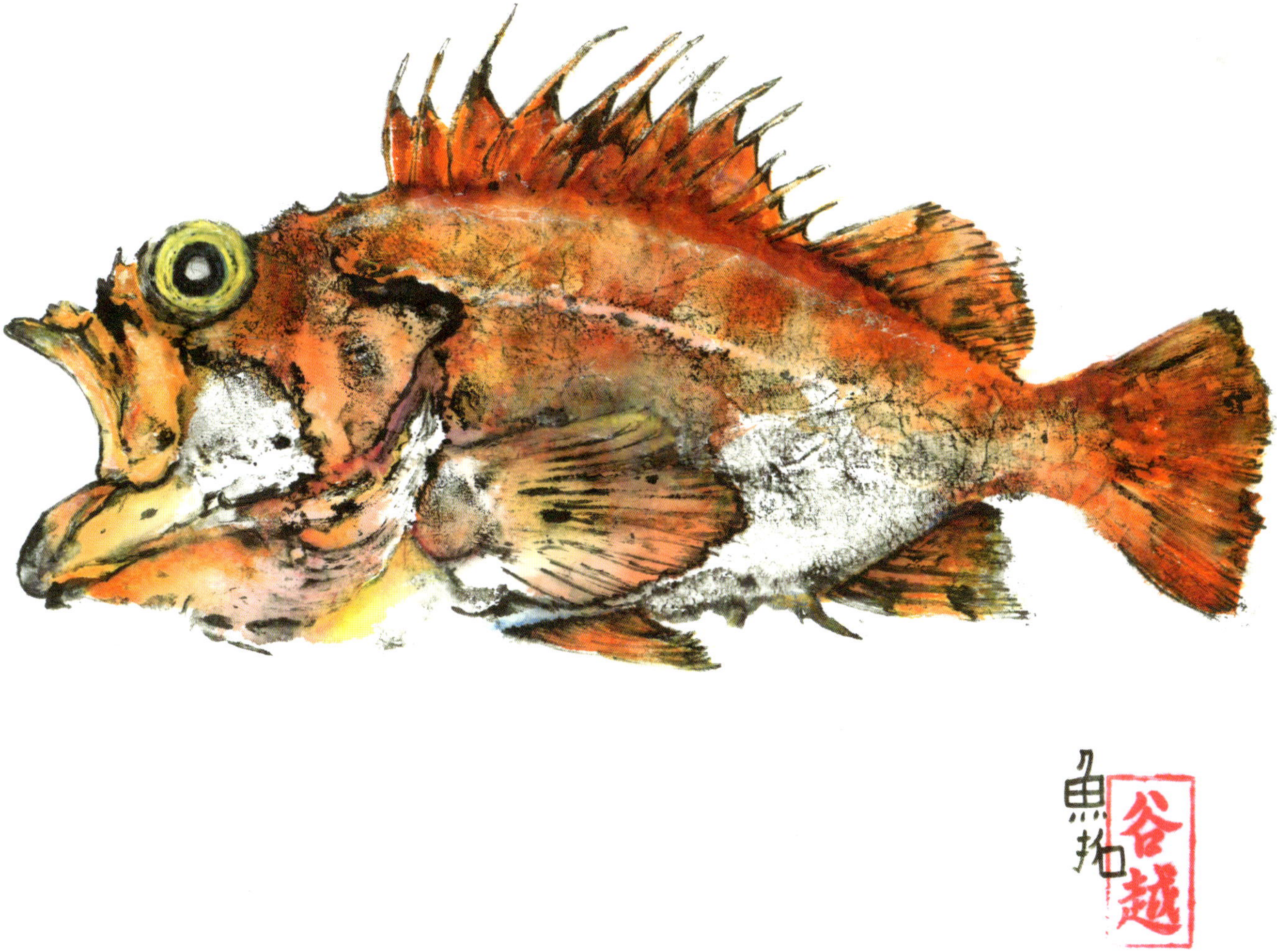

Yelloweye rockfish.

BROWN IRISH LORD SCULPIN

(*Hemilepidotus spinosus*)

COMMON NAME: Bullhead

LIFESPAN: Up to five years

WEIGHT: 1 to 2 pounds

HABITAT: Shallow sandy and rocky habitats

The sculpin is a bottom-dwelling fish. The Brown Irish Lord sculpin's ventral side is brownish-reddish in color, while its dorsal side is entirely white. The Brown Irish Lord sculpin has nostrils that have fleshy flaps and a scaled stripe above the lateral line that is seven- to eight-scales wide. Its pectoral fins are very large. Their lower edge comes out farther than the top edge. Conveniently, the sculpin can simply rest on its pectoral fins.

The Brown Irish Lord sculpin is small, only weighing between 1 and 2 pounds. It can measure up to around 12 inches long.

The Brown Irish Lord sculpin lives in shallow sandy and rocky habitats. It is a nocturnal fish. The way it moves is not necessarily swimming, but a little hop. Interestingly, all fish of the *Cottidae* family lack swim bladders, facilitating their existence on the ocean floor.

Its range is from southern Alaska down to southern California. It can live up to five years.

Because there are so many differences among the juveniles and adults, males and females, it is a daunting task to study the 102 known species of sculpin in Alaska.

The sculpin is commonly called a bullhead because of its disproportionately large head.

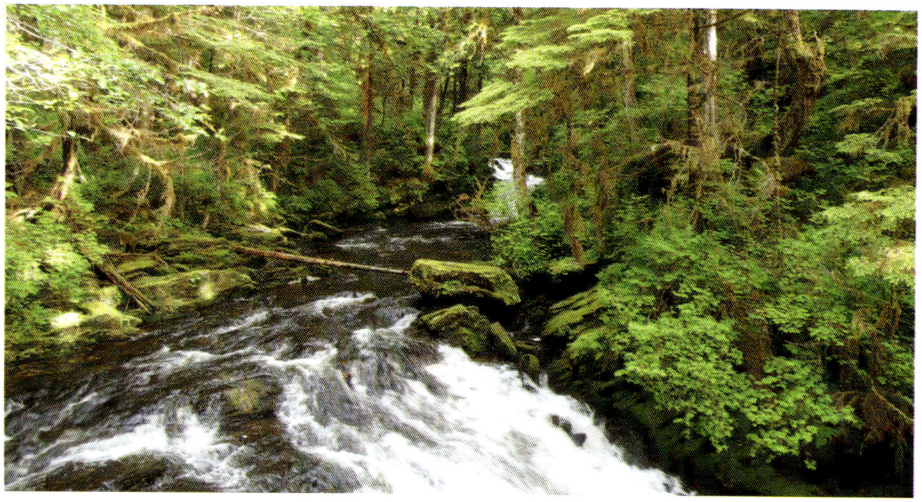

Water racing at Settler's Cove, Ketchikan, Alaska.

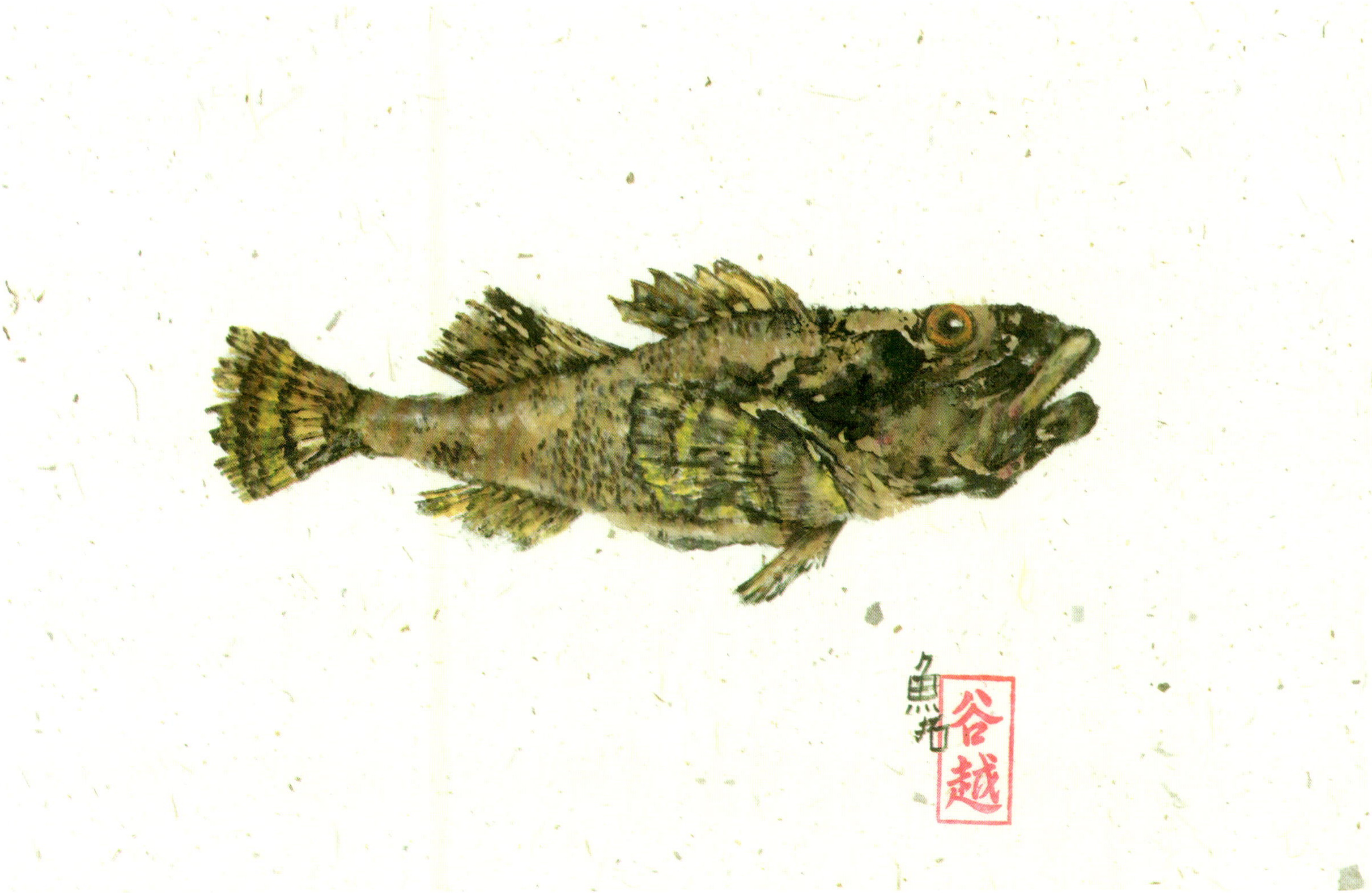

Brown Irish Lord sculpin.

Red Irish Lord Sculpin

(*Hemilepidotus hemilepidotus*)

COMMON NAME: Bullhead

LIFESPAN: Up to six years

WEIGHT: Up to 2.45 pounds

HABITAT: Shallow sandy and rocky areas

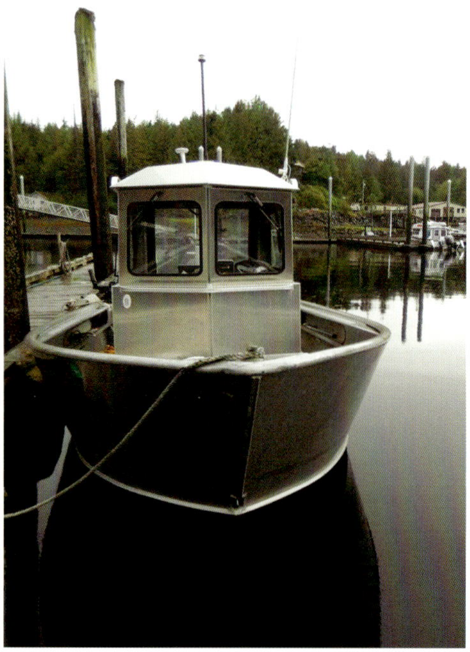

Aluminum fishing boat, Knudson Cove, Ketchikan, Alaska.

The Red Irish Lord is named for its red, white, and brown mottling which gives it a striking appearance. Its mottled colors blend in well with the surrounding rocks and seaweed. It can grow up to 11 inches in length and weigh up to 2.45 pounds.

The Red Irish Lord sculpin has ten to twelve dorsal spines, eighteen to twenty dorsal soft rays, and thirty-five vertebrae. It has four to five rows of scales at the base of its dorsal fin and about ten more scales below the lateral line. It does not have fleshy flaps by its nostrils, as does the Brown Irish Lord sculpin.

The Red Irish Lord sculpin is an ambush predator who waits for its prey to come near and, using its huge mouth, engulfs it. It lives in shallow sandy and rocky areas in saltwater habitats.

The Red Irish Lord sculpin feeds upon mussels, crabs, and barnacles. It is short-lived, with a life span of only up to six years.

The range of the Red Irish Lord sculpin is from the Bering Sea in Alaska to Washington.

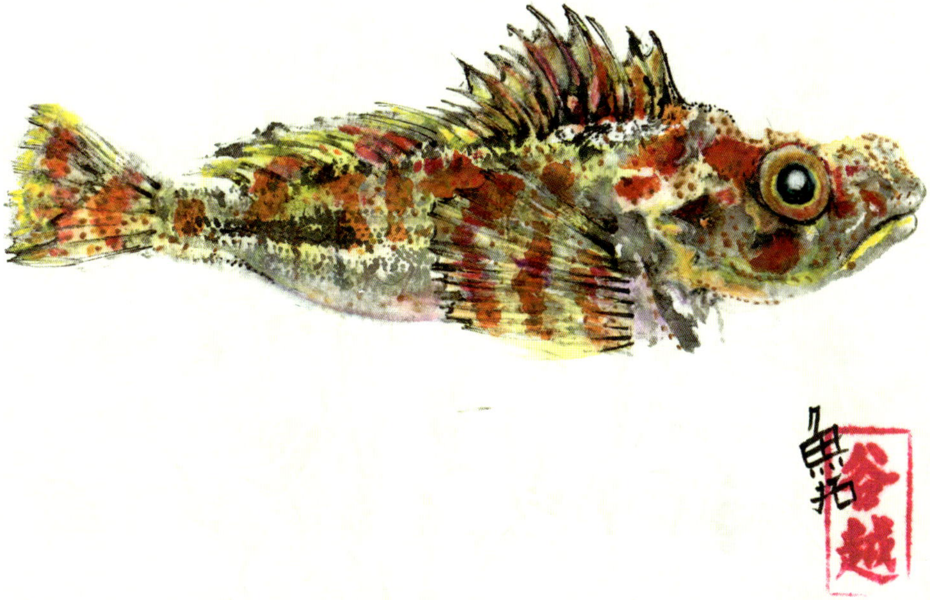

Red Irish Lord sculpin.

SHINER PERCH

(*Cymatogaster aggregata*)

COMMON NAME: Pogie

LIFESPAN: Up to eight years

WEIGHT: Less than half a pound

HABITAT: Shallow waters to 15 meters, but can be found up to 150 meters deep

The shiner perch is a small fish with three yellowish vertical bars on each side of its body. It has large shiny scales, each with a black spot in the center. Breeding males turn a very dark, almost black, color.

Shiner perch are tiny! They usually weigh less than half a pound and only measure up to 8 inches in length.

They are a schooling fish that typically live in water up to 15 meters deep. However, they have been found at depths up to 150 meters. Shiner perch can live up to eight years.

The shiner perch's range is from southern Alaska down to northern Mexico. Typically, shiner perch are caught from piers and docks. It is endearing to watch kids fishing happily for them in the summertime. This would be a great fish for kids to practice gyotaku on!

Native Alaskans commonly call them pogies.

Water off Joe Island, Alaska.

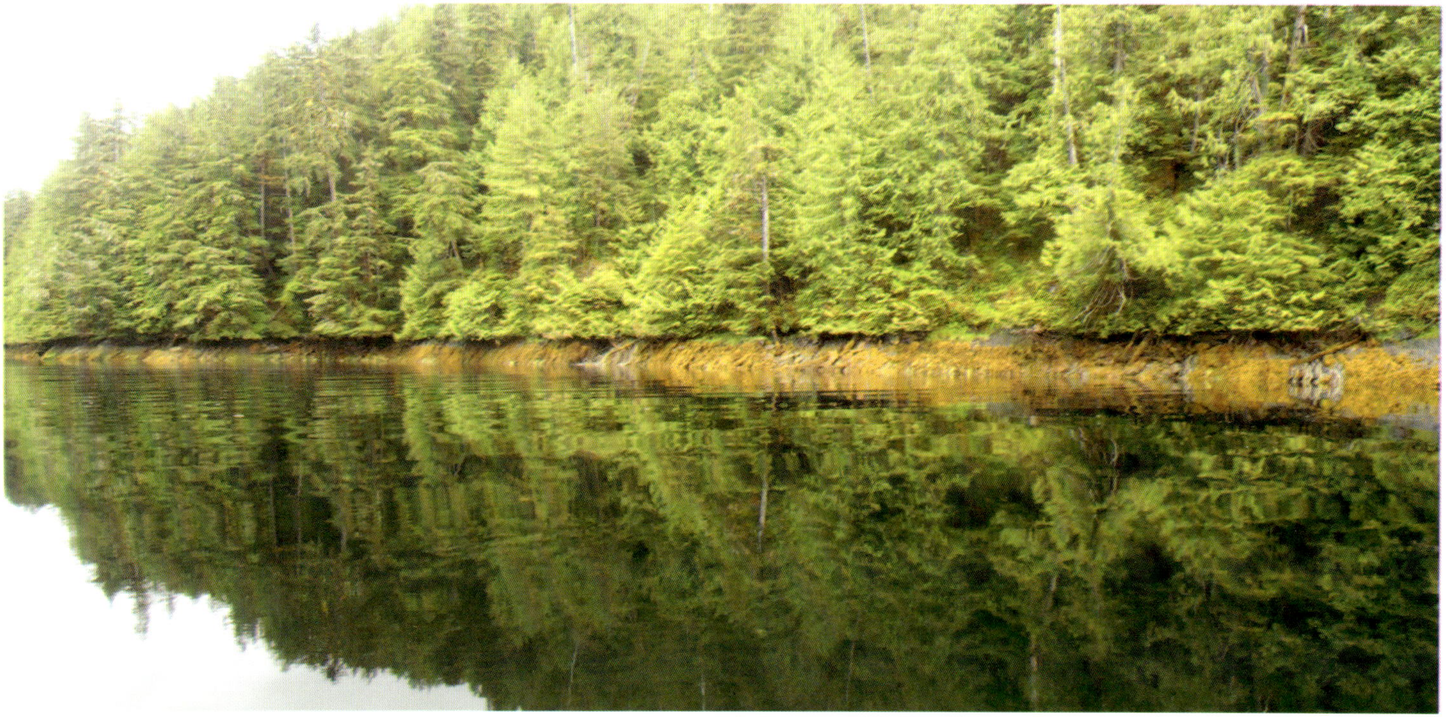

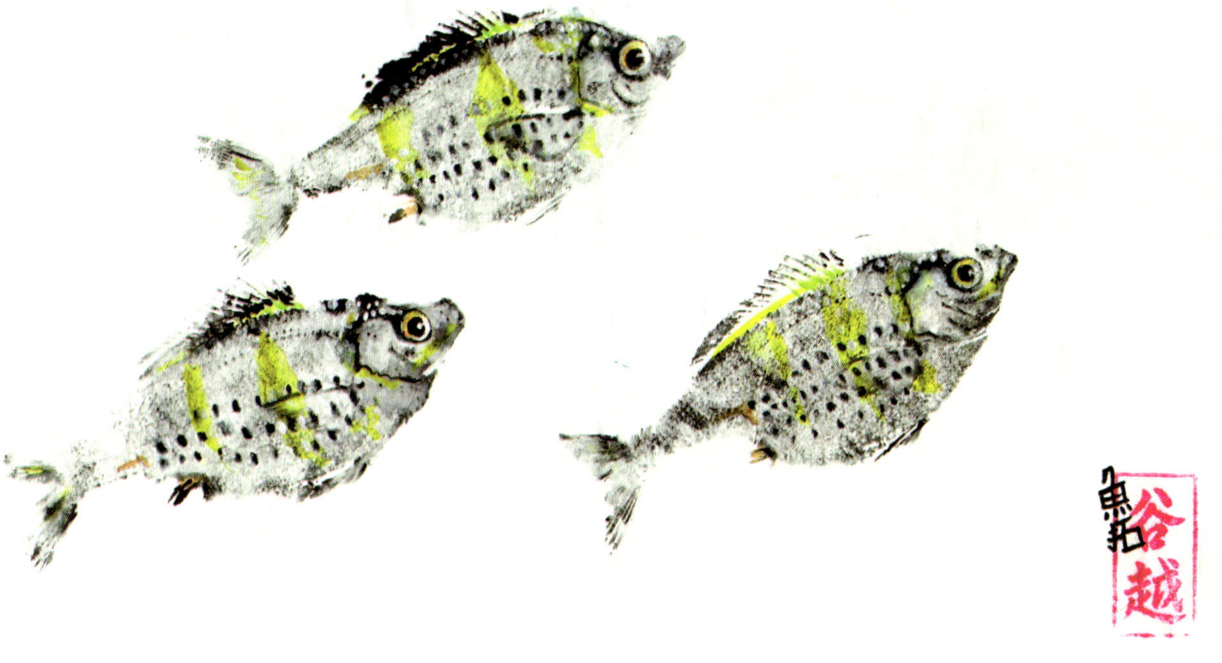

Shiner perch.

Arrowtooth Flounder

(*Atheresthes stomias*)

COMMON NAME: Turbot

LIFESPAN: Up to fifteen years

WEIGHT: Up to 17 pounds

HABITAT: Sandy ocean floor from 18 to 730 meters deep

The arrowtooth flounder is a large flatfish with two rows of very sharp teeth. It has its eyes on the top of its head and swims horizontally. It is brownish green in color on its dorsal side and white on its belly. This flatfish blends in well with the sandy bottom that it calls home.

The arrowtooth flounder is large and can measure up to 34 inches in length and 17 pounds in weight. One of the short-lived fish I've collected, arrowtooth flounder only live for about fifteen years.

It can be found in waters 18 to 730 meters deep. The range of its habitat is from the Bering Sea in Alaska south to Santa Rosa, California.

The arrowtooth flounder is often mistakenly called a turbot. The true turbot (*Scophthalmus maximus*) is also a large flatfish like the arrowtooth flounder. It is prized for its delicate white meat. The true turbot, however, is much rounder in shape than the arrowtooth flounder, and it lacks the large, sharp teeth that the arrowtooth has.

Interestingly, the meat of the arrowtooth flounder is considered inedible. This is due to a proteolytic enzyme that is given off by a myxosporean parasite when the meat is heated. This renders the meat too mushy to eat.

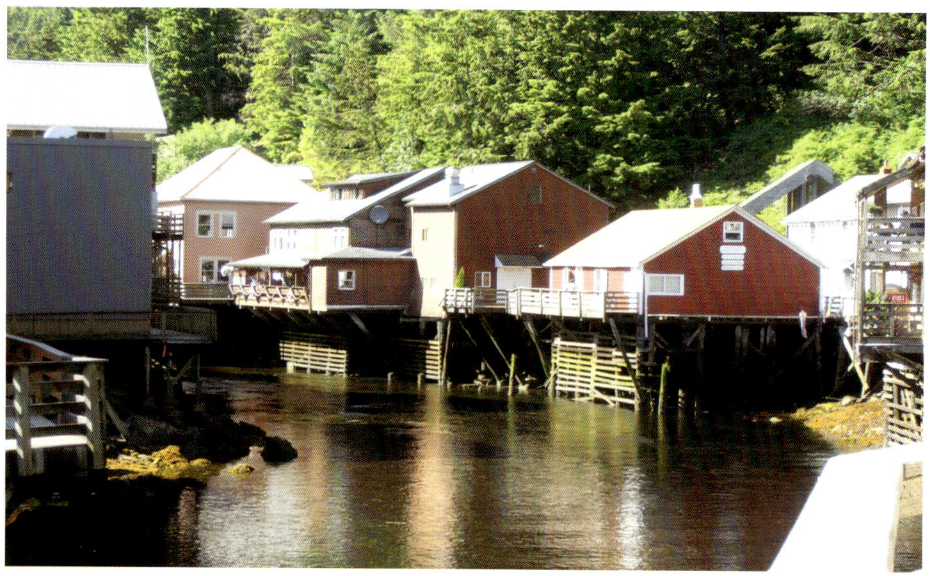

Creek Street, Ketchikan, Alaska.

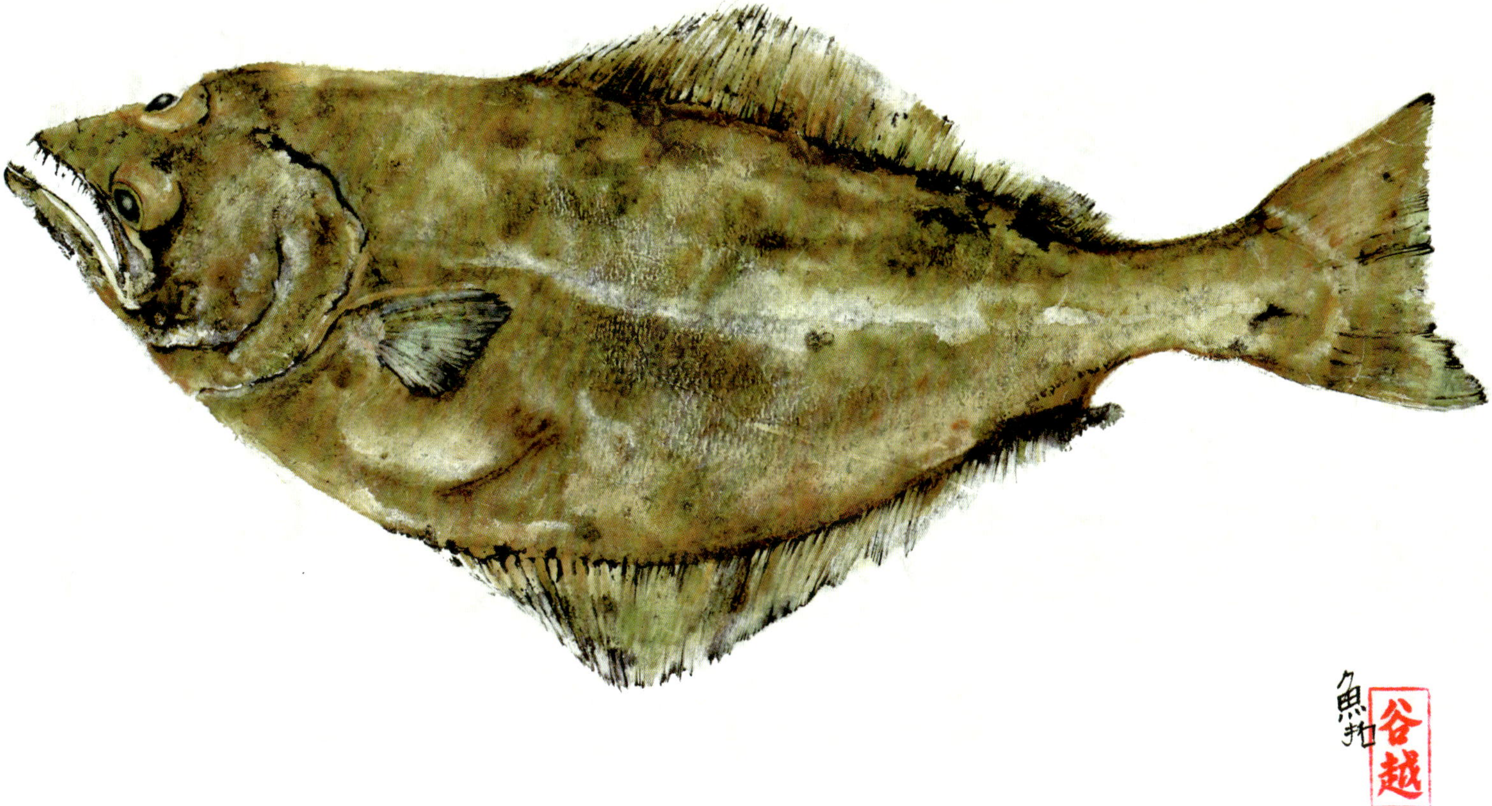

Arrowtooth flounder.

ROCK SOLE

(*Lepidopsetta bilineata*)

COMMON NAMES: Broadfin sole, roughscale sole, two-lined flounder

LIFESPAN: Up to twenty-six years

WEIGHT: Up to 4 pounds

HABITAT: Sandy, muddy, and gravelly bottoms up to 576 meters deep

The rock sole is a member of the "ray-finned" flatfishes, along with its relatives the starry flounder, halibut, and turbot, that begin their lives as juveniles who swim upright. The rock sole is known as a right-eyed fish, as is the arrowhead turbot, because both of its eyes are on its right side. Most flatfish are left-eyed. With time, the rock sole begins to tilt and its eyes start to migrate so that it eventually is horizontal and flat, hence its categorization as a flatfish. The rock sole has dark blotches or bars on its dorsal and anal fins.

It can weigh up to 4 pounds and measure up to 24 inches in length. The rock sole can live up to twenty-six years.

The rock sole lives in saltwater on sandy, muddy, and gravelly bottoms from zero to 576 meters deep. Its range is wide, from the Seas of Japan and Okhotsk to the Bering Strait and south to Tanner Bank off the coast of southern California.

In Greek, its scientific name *lepis* means "scale" and *psetta* means "grouper." Its common names include the broadfin sole, the roughscale sole, and the two-lined flounder.

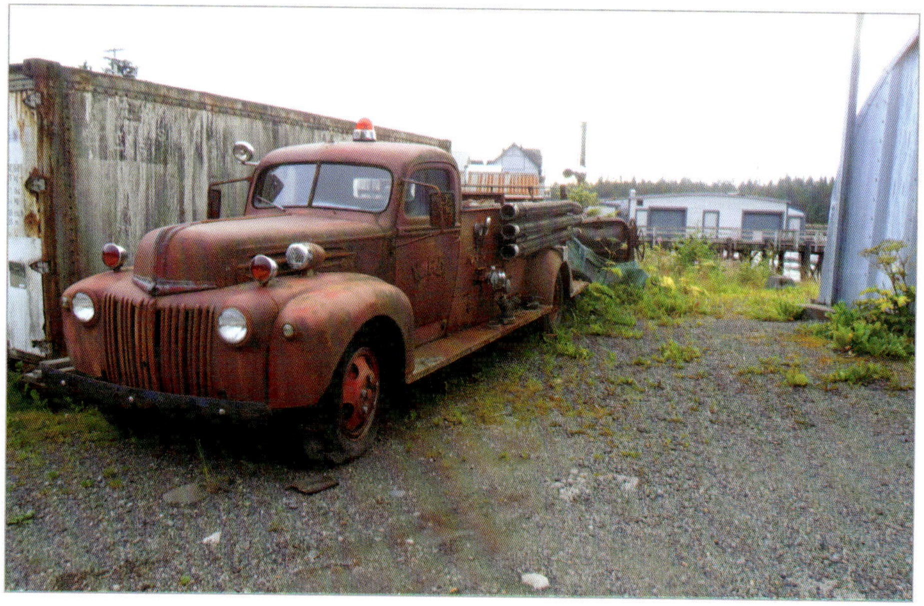

Fire truck on Prince of Wales Island, Alaska.

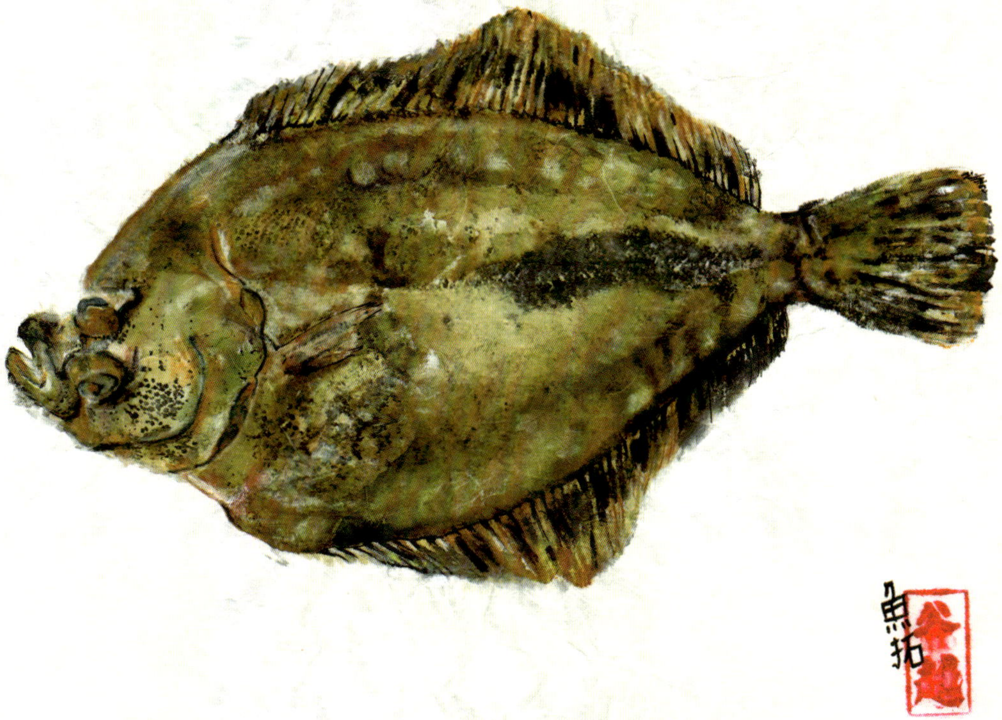

Rock sole.

KELP GREENLING

(*Hexagrammos decagrammus*)

COMMON NAMES: Greenling sea trout, rock trout, spotted rock trout, and kelp trout or cod

LIFESPAN: Up to eighteen years

WEIGHT: Up to 4.5 pounds

HABITAT: Kelp beds and sandy ocean floor up to 45 meters deep

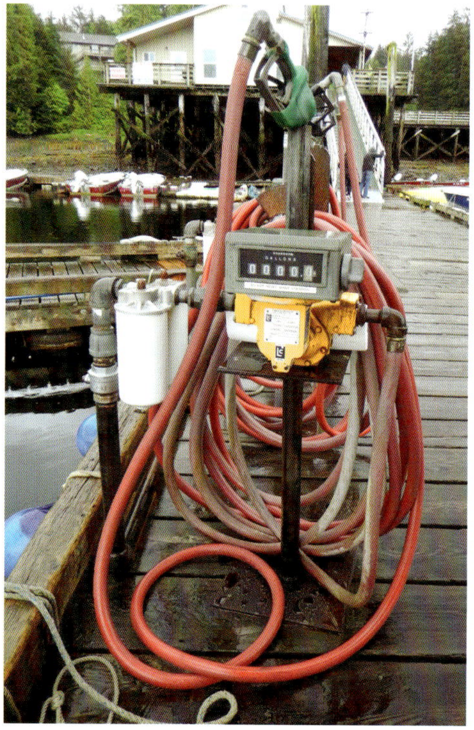

Fuel pump at Knudson Cove, Ketchikan, Alaska.

The female and male kelp greenlings have distinctive appearances. The female of the species is light brown in color with bright-yellow spots, while the male is grayish or olive-colored with bright-blue spots. Its dorsal fin has no sharp spines.

Like the sculpin, the kelp greenling lacks a swim bladder. Its body is fusiform and adapted to swimming in open water. The pectoral fins of the greenling are disproportionately large.

The kelp greenling can weigh up to around 4.5 pounds and measure up to 21 inches. They can live to be eighteen years old. The kelp greenling habitat ranges from the Aleutian Islands, Alaska, down south to La Jolla, California.

Kelp greenling live in saltwater at depths to 45 meters. They live around kelp beds and on the sandy bottoms of the ocean floor. Kelp greenling can also live in rocky inshore habitats. The diet of the kelp greenling consists of small fish, brittle stars, mollusks, crustaceans, and polychaete worms.

After the female kelp greenling reaches four years of age, she can participate in multiple spawning events throughout the year. The male kelp greenling protects the sticky clusters of eggs that are laid on rocky outcroppings and kelp. Many females can lay eggs in a single nest.

Common names for the kelp greenling are the greenling sea trout, rock trout, spotted rock trout, and the kelp trout or cod.

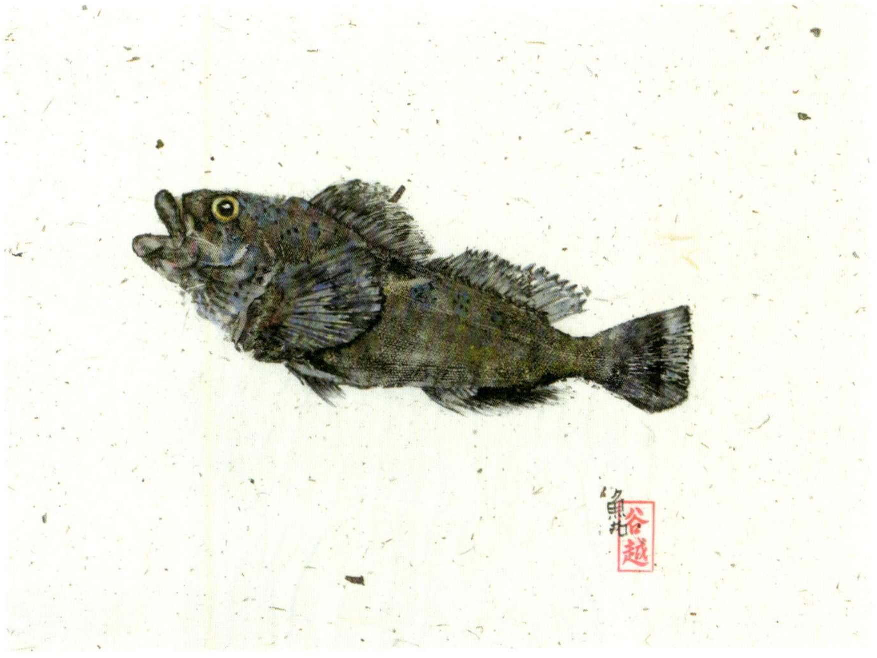

Male kelp greenling.

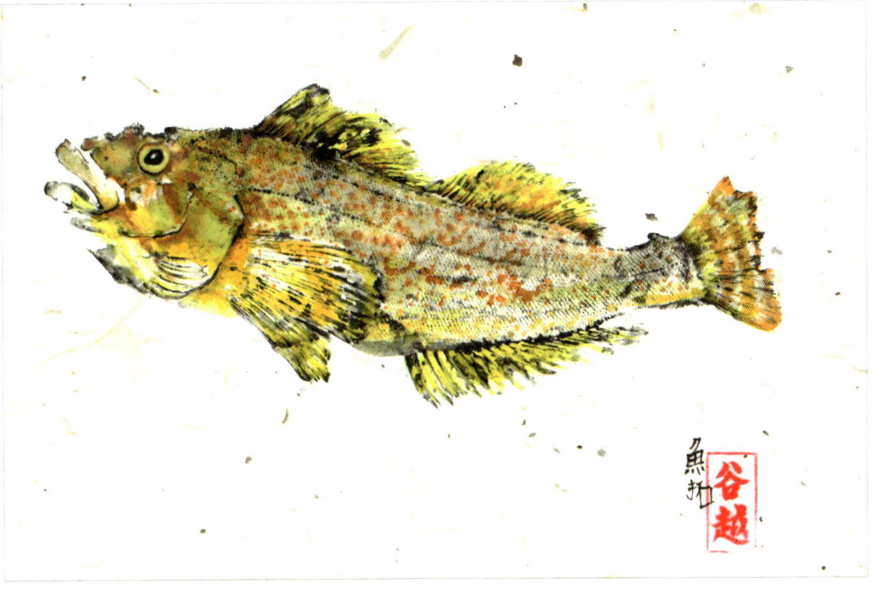

Female kelp greenling.

DUNGENESS CRAB

(Cancer magister)

COMMON NAME: Dungy

LIFESPAN: Up to ten years

WEIGHT: Up to 5 pounds

HABITAT: Ocean floor from shallow water to 610 meters deep

The Dungeness crab was named after a small fishing village in Washington. It is typically brownish orange in color and can camouflage with the sandy bottom of the ocean where it lives. If this decapod feels in danger, it can use its sharp claws to dig into the sand and hide its whole body.

The Dungeness crab can live up to ten years. It can weigh up to 5 pounds and measure up to almost 8 inches across its carapace.

Its range is from Point Conception, California, north to the Aleutian Islands, Alaska. The Dungeness crab is typically found in shallow water, but can be found up to 610 meters deep. It eats clams, crustaceans, and small fish. Only the male crab can be harvested. The females must be returned to the ocean to conserve the species.

The Dungeness crab is also known as a Dungy.

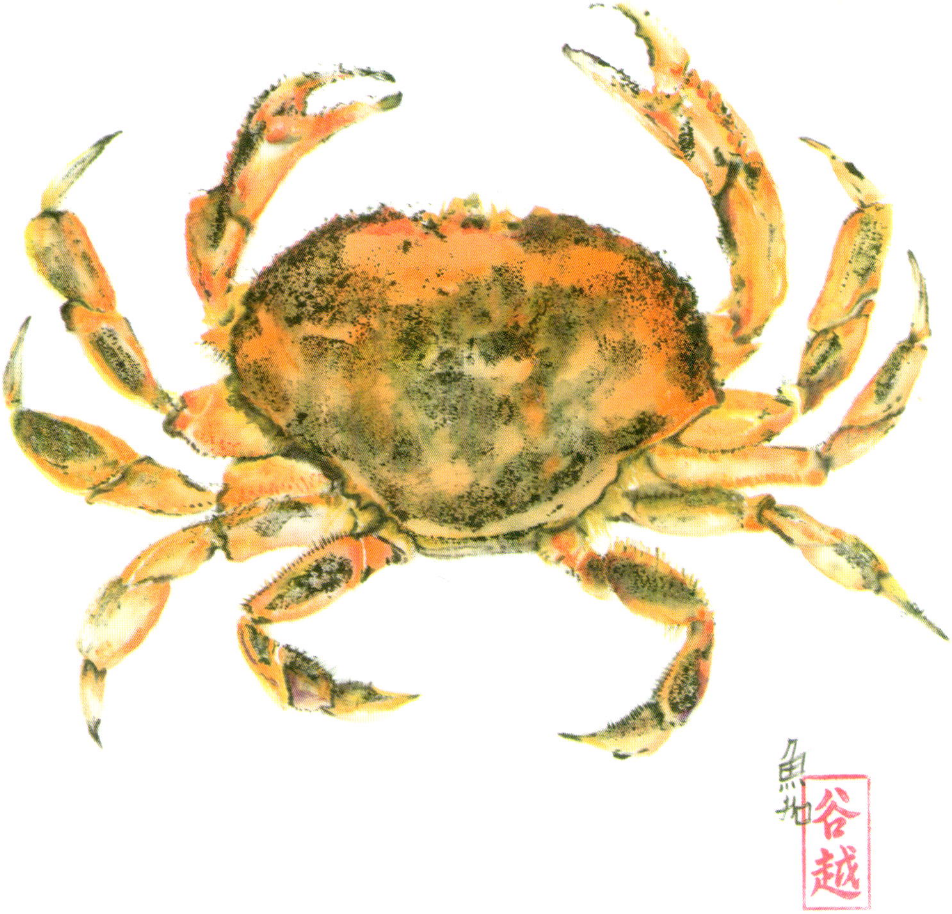

Dungeness crab.

SPOT SHRIMP

(*Pandalus platyceros*)

COMMON NAMES: Alaskan prawn, two-spot prawn

LIFESPAN: Up to eleven years

WEIGHT: Up to one-third of a pound

HABITAT: Rock piles and crevices from 4 to 457 meters deep

The spot shrimp is a beautiful shrimp with a distinctive spot on each side of its body from which it derives its name. The spot shrimp is large and can be over a foot long and weigh up to one-third of a pound.

The spot shrimp live in saltwater around rock piles and in crevices from 4 to 457 meters deep.

Interestingly, members of this genus are protandric hermaphrodites, starting life as males and later becoming females. Additionally, a spot shrimp spawns once as a male and one or more times as a female. The spot shrimp spawns at depths of 500 to 700 feet. The eggs are attached to the female's abdomen by hair-like structures called setae. Eggs are found on females from October to March.

These large shrimp have a lifespan of about eleven years.

If you are looking for other names of the spot shrimp, they also go by Alaskan prawn and the two-spot prawn.

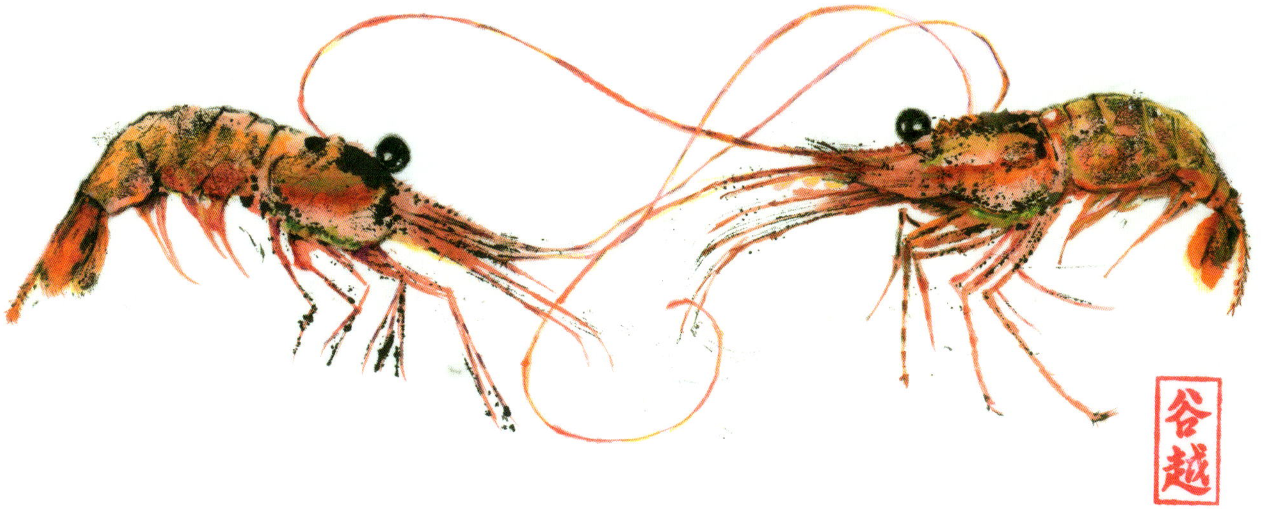

Spot shrimp.

PURPLE SEASTAR

(*Pisaster ochraceus*)

COMMON NAMES: Purple starfish, ochre starfish

LIFESPAN: Four to twenty years

WEIGHT: Up to 1 pound

HABITAT: Tidal zones and rocky areas up to 90 meters deep

As the name suggests, the purple sea star is usually purple, but some can be orange, yellow, reddish, or brown. This sea star has five rays ranging from 5 to 10 inches in length and can weigh up to a pound.

The dorsal side of its body is covered with small spines called ossicles. The ventral side of its body has many tubular feet with suckers, which allow it to be mobile on the tide pool rocks it calls home.

The purple sea star, or starfish, is a keystone species found from Prince William Sound in Alaska down to Santa Barbara, California. It is called a keystone species because its presence or absence has great bearing on the health of the intertidal community.

The purple sea star lives in saltwater tidal zones and rocky areas up to 90 meters deep. The purple sea star's diet consists of mussels, snails, barnacles, limpets, crustacea, and chitons. The purple sea star, like all starfish, has no brain. It relies upon a nerve ring system around the mouth and five radial nerves that control the five legs.

The purple sea star has a lifespan of four to twenty years and usually breeds at around five years old. The purple sea star reproduces by broadcast spawning, in which the female releases eggs into the water column and the male releases sperm that fertilizes the eggs. Unbelievably, a female can produce up to 40 million eggs at one time!

Common names for the purple sea star are the ochre starfish and the purple star fish.

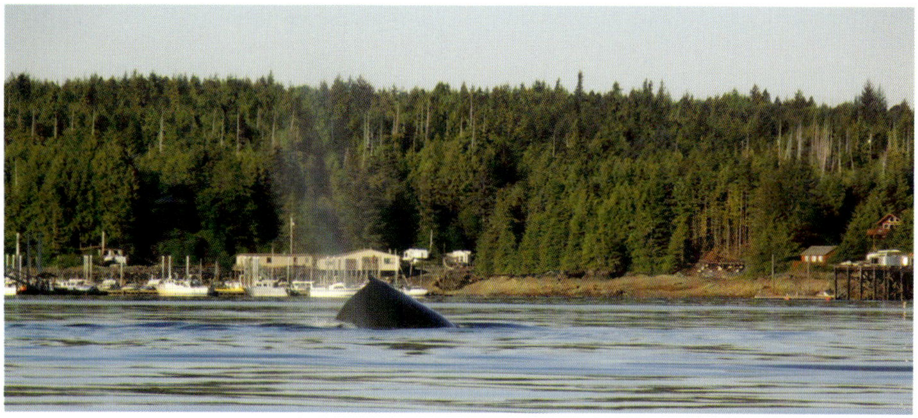

Humpback whale, Knudson Cove, Ketchikan, Alaska.

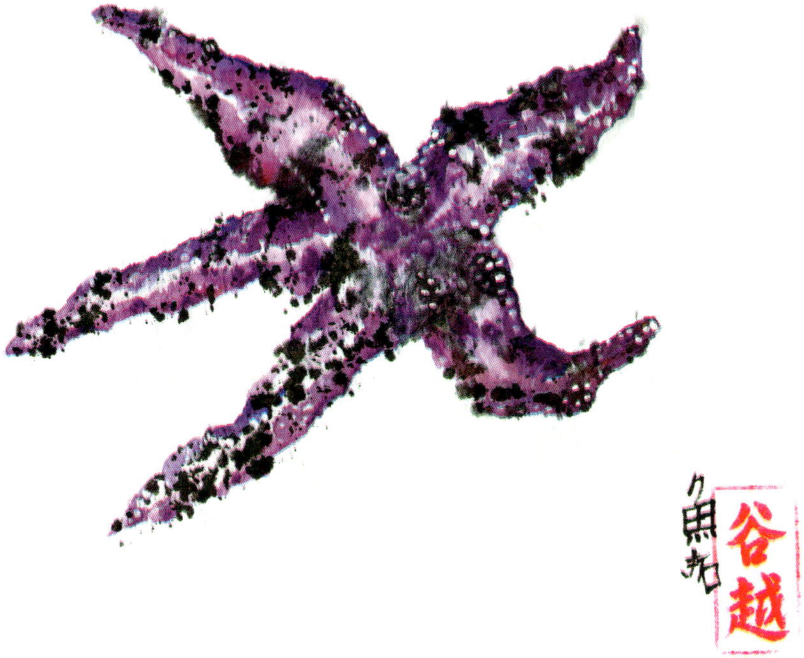

Purple sea star.

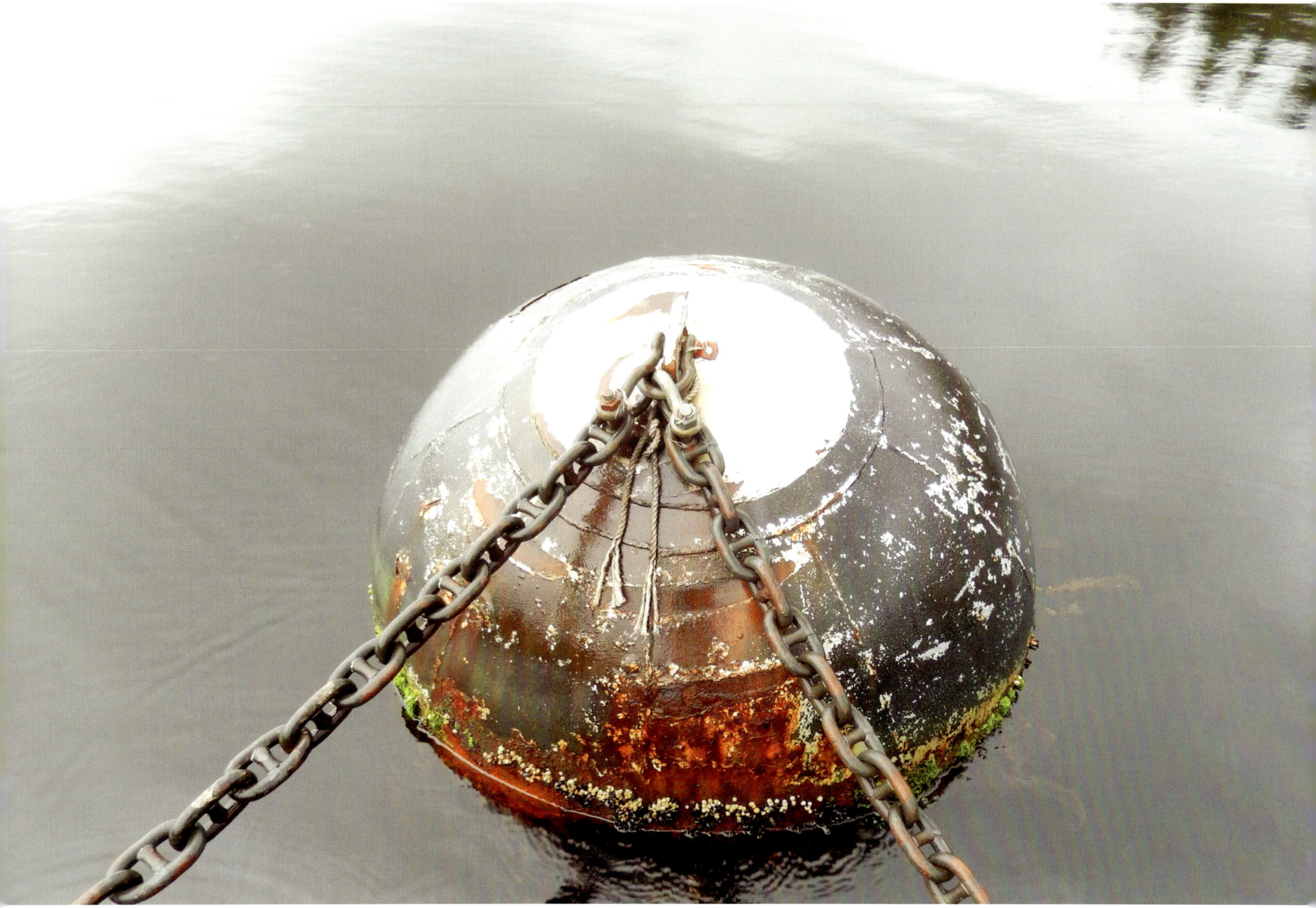

Mooring buoy off dock.

Conclusion

In hindsight, it was my curiosity and adoration of the ancient Japanese art of gyotaku that led me to embark on this adventure of chronicling the diversity of Southeast Alaska's marine life. Every day of my adventure opened my eyes to marine life that I could only have imagined existed. Armed with a paintbrush, sumi ink, watercolors and a camera, I set off to write and illustrate this book. I want to take this opportunity to thank you for sharing this adventure with me, and, as I conclude, please allow me to pose my opening question to you again:
"Do you gyotaku?"
Very clearly . . . yes, you do!

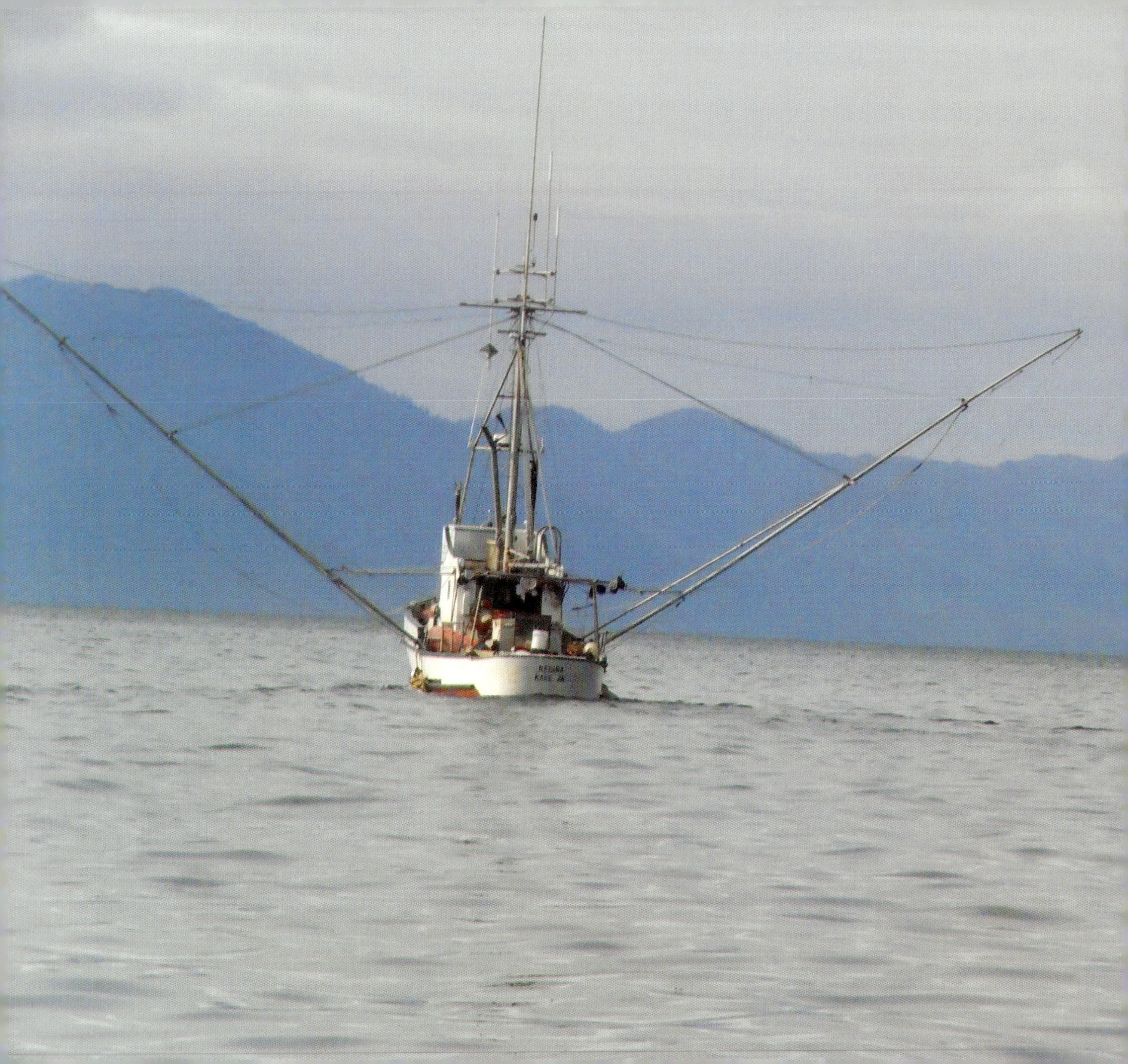

BIBLIOGRAPHY

Alaska Department of Fish and Game. "Black Rockfish (*Sebastes melanops*) Species Profile." Accessed February 16, 2018. http://www.adfg.alaska.gov/index.cfm?adfg=blackrockfish.main.

Alaska Department of Fish and Game. "Dungeness Crab Species Profile." Accessed February 16, 2018. www.adfg.alaska.gov/index.cfm?adfg=dungenesscrab.main.

Alaska Department of Fish and Game. "Pacific Salmon Marine Phase Identification." January 2014. Accessed February 16, 2018. https://www.adfg.alaska.gov/static/fishing/pdfs/sport/SalmonID2014.pdf

Alaska Department of Fish and Game. "Pelagic and Non-Pelagic Rockfish Change the Way You Fish." Accessed February 16, 2018. https://www.adfg.alaska.gov/static/regularions/fishregulations/PDFs/pdfs/rockfishID.pdf.

Alaska Department of Fish and Game. "Pink Salmon (*Oncorhynchus gorbuscha*) Species Profile." Accessed February 16, 2018. http://www.adfg.alaska.gov/index.cfm?adfg=pinksalmon.main.

Alaska Department of Fish and Game. "Rockfish through the Ages." Alaska Fish & Wildlife News. April 2008. Accessed April 4, 2018. http://www.adfg.alaska.gov/index.cfm?adfg=wildlifenews.view_article&articles_id=366.

Alaska Department of Fish and Game. "Spot Shrimp Species Profile." Accessed February 16, 2018. www.adfg.alaska.gov/index.cfm?adfg=spotshrimp.main.

Long-liner off Grant Island, Alaska.

Alaska Department of Fish and Game. "Yelloweye Rockfish (*Sebastes ruberrimus*) Species Profile." Accessed February 16, 2018. http://www.adfg.alaska.gov/index.cfm?adfg=yelloweyerockfish.main.

Alaska Fisheries Science Center. "Arrowtooth Flounder Research." Accessed February 16, 2018. https://www.afsc.noaa.gov/species/Arrowtooth_flounder.php.

Alaska Fisheries Science Center. "Black Rockfish." Accessed February 16, 2018. https://www.afsc.noaa.gov/groundfish/RockfishGuide/Rockfish_Pages/Black_rockfish.htm.

Alaska Fisheries Science Center. "Black Rockfish, *Sebastes melanops*." Accessed February 16, 2018. https://www.afsc.noaa.gov/kodiak/photo/fishblkrock.htm.

Alaska Fisheries Science Center. "Copper Rockfish." Accessed February 16, 2018. https://www.afsc.noaa.gov/Rockfish-Game/description/copper.htm.

Alaska Fisheries Science Center. "Dark Dusky Rockfish." Accessed February 16, 2018. https://www.afsc.noaa.gov/groundfish/RockfishGuide/Rockfish_Pages/Dark_dusky_rockfish.htm.

Alaska Fisheries Science Center. "Pacific Ocean Perch." Accessed March 3, 2018. https://www.afsc.noaa.gov/Education/factsheets/10_POP_fs.pdf.

Alaska Fisheries Science Center. "Quillback Rockfish." Accessed February 16, 2018. https://www.afsc.noaa.gov/groundfish/RockfishGuide/Rockfish_Pages/Quillback_rockfish.htm.

Alaska Fisheries Science Center. "Rockfish Species List." Accessed February 16, 2018. https://www.afsc.noaa.gov/groundfish/RockfishGuide/species_list.htm.

Alaska Fisheries Science Center. "Silvergray Rockfish." Accessed February 16, 2018. https://www.afsc.noaa.gov/rockfish-game/description/silvergray.htm.

Alaska Fisheries Science Center. "Starry Founder [*sic*], *Platichthys stellatus*." Accessed February 16, 2018. https://www.afsc.noaa.gov/kodiak/photo/fishstarry.htm.

Alaska Fisheries Science Center. "Tiger Rockfish." Accessed February 16, 2018. https://www.afsc.noaa.gov/groundfish/RockfishGuide/Rockfish_Pages/Tiger_rockfish.htm.

Ancient Origins. "Gyotaku: A Unique Japanese Tradition in Which a Fish Becomes a Work of Art." Accessed February 16, 2018. http://www.ancient-origins.net/history-ancient-traditions/gyotaku-unique-japanese-tradition-which-fish-becomes-work-art-004390?nopaging=1.

Barefield, Robin. "King (Chinook) Salmon (*Oncorhynchus tshawytscha*)." November 30, 2017. Accessed February 16, 2018. http://robinbarefield.com/king-chinook-salmon-oncorhynchus-tshawytscha/.

California Department of Fish and Wildlife. "Anglers Reminded to Avoid Yelloweye Rockfish." CDFW News. August 25, 2017. Accessed March 3, 2018. https://cdfgnews.wordpress.com/2017/08/25/anglers-reminded-to-avoid-yelloweye-rockfish/.

Chilton, E. A. "Maturity and Growth of Female Dusky Rockfish (*Sebastes variabilis*) in the Central Gulf of Alaska." Kodiak, Alaska, pp. 70–78. NOAA Fisheries. 2010. Accessed February 16, 2018. https://www.st.nmfs.noaa.gov/spo/FishBull/1081/chilton.pdf.

Danovich, Tove. "How Fishermen's Bragging Rights Gave Birth to Fine Art." The Salt, NPR. August 30, 2015. Accessed March 3, 2018. https://www.npr.org/sections/thesalt/2015/08/30/435248291/how-fishermen-s-bragging-rights-gave-birth-to-fine-art.

Dodge, K. Erica. "Gyotaku: The Ancient Japanese Art of Printing Fish." TedEd. Accessed March 2, 2018. https://ed.ted.com/lessons/gyotaku-the-ancient-japanese-art-of-printing-fish-k-erica-dodge.

The Editors of *Encyclopædia Britannica*. "Sculpin." *Encyclopædia Britannica*. March 24, 2017. Accessed February 16, 2018. https://www.britannica.com/animal/sculpin.

Fish-Fishes. "Starry Flounder (Lat. *Platichthys stellatus*)." Accessed February 16, 2018. https://www.fish-fishes.com/salt-water-fish/starry-flounder-lat-platichthys-stellatus/.

Fretwell, Kelly, and Brian Starzomski. "Shiner Perch, Shiner Surfperch, Yellow Shiner, *Cymatogaster aggregata*." Biodiversity of the Central Coast. 2014. Accessed February 16, 2018. http://www.centralcoastbiodiversity.org/shiner-perch-bull-cymatogaster-aggregata.html.

Fukuchi, Mitsuo, Harvey J. Marchant, and Boshu Nagase. *Antarctic Fishes*. Baltimore: Johns Hopkins University Press, 2007.

Gumerson, Jessica. "*Sebastes caurinus* (Rockfish)." Animal Diversity Web. Accessed February 16, 2018. http://animaldiversity.org/accounts/Sebastes_caurinus/.

Halstead, Bruce W., and Edward T. Roche. "The Venom Apparatus of California Rockfishes (Family *Scorpaenidae*)," State of California, the Resources Agency, Department of Fish and Game Fish Bulletin 156. 1972. Accessed March 3, 2018. http://content.cdlib.org/view?docId=kt8z09n9rg&brand=calisphere&doc.view=entire_text.

History. "Tokugawa Period and Meiji Restoration." Accessed March 2, 2018. http://www.history.com/topics/meiji-restoration.

Hochhalter, Sam. "Amazing Rockfish New Release Technique Greatly Increases Survival." Alaska Department of Fish and Game. Accessed March 3, 2018. https://www.adfg.alaska.gov/index.cfm?adfg=wildlifenews.view_article&articles_id=530.

Its Nature. "Pink Salmon." Accessed February 16, 2018. https://itsnature.org/sea/fish/pink-salmon/.

Josey, Terry, and Donald E. Kramer. *Guide to Northeast Pacific Flatfishes: Families* Bothidae, Cynoglossidae, *and* Pleuronectidae. Anchorage: Alaska Fisheries Development Foundation, 2008.

Kachemak Bay Research Reserve. "Common Sculpin of Cook Inlet." Alaska Center for Conservation Science. 2012. Accessed February 16, 2018. http://accs.uaa.alaska.edu/files/kachemak-bay/commonsculpincookinlet.pdf.

Kendall, Penny. "Side Effects of Stepping on Rock Fish." Gone Outdoors. Accessed March 3, 2018. https://goneoutdoors.com/side-effects-stepping-rock-fish-6889957.html.

Lewin, Joe. "The Health Benefits of Salmon." July 10, 2017. Accessed March 1, 2018. https://www.bbcgoodfood.com/howto/guide/ingredient-focus-salmon.

Love, Milton S., Lyman Thorsteinson, and Mary Yoklavich. *The Rockfishes of the Northeast Pacific*. Berkeley: University of California Press, 2002.

Munk, Kristen M. "Maximum Ages of Groundfishes in Waters off Alaska and British Columbia and Considerations of Age Determination." Alaska Department of Fish and Game. Accessed May 23, 2018. https://www.adfg.alaska.gov/static/home/library/PDFs/afrb/munkv8n1.pdf.

National Wildlife Federation. "Chinook Salmon." Accessed February 16, 2018. https://www.nwf.org/en/Educational-Resources/Wildlife-Guide/Fish/Chinook-Salmon.

NOAA Fisheries. "Chinook Salmon (*Oncorhynchus tshawytscha*)." January 21, 2015. Accessed February 16, 2018. http://www.fisheries.noaa.gov/species/chinook-salmon.html.

NOAA Fisheries. "Coho Salmon (*Oncorhynchus kisutch*)." January 21, 2015. Accessed February 16, 2018. http://www.nmfs.noaa.gov/pr/species/fish/coho-salmon.html.

NOAA Fisheries. "Yelloweye Rockfish (*Sebastes ruberrimus*)." January 2, 2015. Accessed February 16, 2018. https://www.fisheries.noaa.gov/species/yelloweye-rockfish.html.

NOAA Fisheries Alaska Regional Office. "Nearshore Fish Photos." Accessed February 16, 2018. https://alaskafisheries.noaa.gov/habitat/nearshore-fish-photos.

Oddity Central. "Gyotaku—The Traditional Japanese Art of Painting Fish with Actual Fish." September 9, 2015. Accessed February 16, 2018. http://www.odditycentral.com/pics/gyotaku-the-traditional-japanese-art-of-painting-fish-with-actual-fish.html.

"*Platichthys stellatus*." *Platichthys stellatus* (Starry Flounder). Accessed March 3, 2018. https://wdfw.wa.gov/fishing/bottomfish/identification/flatfish/p_stellatus.html.

Ramirez, Yesenia. "*Pisaster ochraceus*." Animal Diversity Web. Accessed February 16, 2018. http://animaldiversity.org/accounts/Pisaster_ochraceus/.

The Safina Center. "Spot Shrimp—ALASKA." Accessed February 16, 2018. safinacenter.org/documents/2012/03/shrimp-spot-alaska-full-species-report.pdf.

ThisFish. "Yelloweye Rockfish." 2013. Accessed February 16, 2018. https://thisfish.info/fishery/species/yelloweye-rockfish/.

Washington Department of Fish and Wildlife. "Bottomfish Identification." Accessed February 16, 2018. https://wdfw.wa.gov/fishing/bottomfish/identification/rockfish/s_melonops.html.

Washington Department of Fish and Wildlife. "Bottomfish Identification Guide: Greenling." Accessed February 16, 2018. https://wdfw.wa.gov/fishing/bottomfish/identification/greenling/h_decagrammus.html.

Washington Department of Fish and Wildlife. "Brown Irish Lord." Accessed February 16, 2018. https://wdfw.wa.gov/fishing/bottomfish/identification/sculpins/h_spinosus.html.

Washington Department of Fish and Wildlife. "Canary Rockfish." Accessed February 16, 2018. https://wdfw.wa.gov/fishing/bottomfish/identification/rockfish/s_pinniger.html.

Washington Department of Fish and Wildlife. "Tiger Rockfish." Accessed February 16, 2018. https://wdfw.wa.gov/fishing/bottomfish/identification/rockfish/s_nigrocinctus.html.

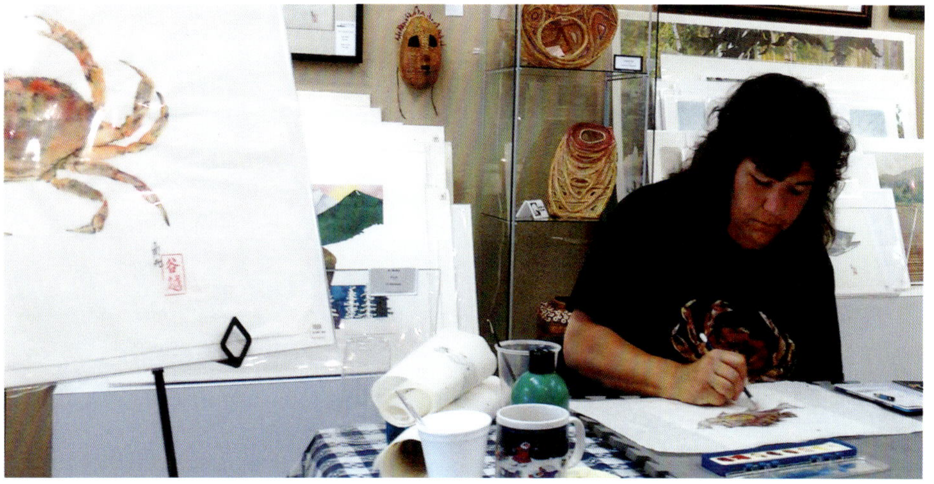

Photo by Carlene Tinker.

JULIA TANIGOSHI TINKER is an artist who works with mixed media, combining the traditional Japanese art form of gyotaku, 魚拓, with watercolor to create unique paintings of fish and crustaceans from southeast Alaska. This art form, in which a print is made of a fish using non-toxic sumi ink on washi paper, originated in Japan in the early 1800s as a way for fishermen to remember their prized catches. Julia Tanigoshi Tinker has exhibited her paintings at shows both in California and in Ketchikan, Alaska, and has received various awards for them. She loves to use watercolor techniques she studied at the Instituto de Bellas Artes in San Miguel de Allende, Mexico, along with her passion for fishing, to create one-of-a-kind gyotaku paintings celebrating the diversity of southeast Alaska's marine ecosystems.

When she is not fishing and painting, she is busy teaching Spanish in high school to ninth through twelfth graders. She loves the energy and enthusiasm they bring. She enjoys telling stories of her adventures and sharing her creativity with her students.

An avid outdoorswoman, she spends her summers fishing throughout southeast Alaska. She spent the last few years preparing her paintings and photographs in this book Gyotaku prints of southeast Alaska's fish and crustaceans. She caught each of these specimens while captaining her own boat and can tell a story about each one.

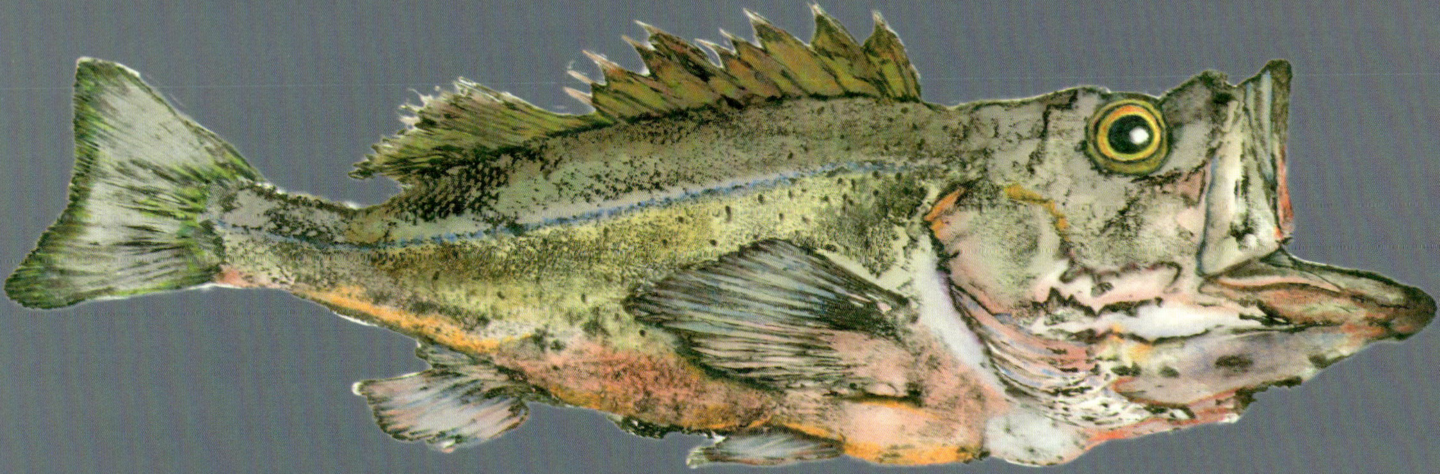

Sea Bass.